FAITH IN LIFE

LIFE ITSELF AS A MODERN RELIGION

Book 3 – Additional Thoughts

By Charles Blaise

TO REASON PUBLISHING

Library of Congress Control Number: 2005909343
ISBN 978-0-9774206-8-1

Printed in the United States of America

To Reason Publishing/www.toreasonpublishing.com
"Life According To Reason."– Aristotle
P. O. Box 237037
New York, NY 10023

CONTENTS

FOREWORD

Book Three presents additional thoughts on *Life Itself As A Modern Religion.* Since the religion is based on what we can credibly know, the beliefs exist beyond the usual disputes between science and religion. For example, the fundamental belief is Faith In Life, which is founded on an appreciation of the greatness of life as we have come to understand it. We can actualize our faith with intelligent Devotion To Life, a precept that makes our lives consonant with the purpose that is evident in Creation's own overall devotion to evolving and sustaining life within what we perceive as natural variability.

I suggest a scientifically credible definition of what we, as informed and rational moderns, can consider God, and provide, as the foundation of a practical ethic, the seemingly daring concept that God is life, as a naturally miraculous attribute of the universe, taken as our Natural Supreme Being, or as the animate representative of any Creator or Creators who may be outside of our experience.

While there is some overlap between the ideas presented in Book One and Book Three, I have done what I can to limit it while allowing for the expression of the same thoughts in different but individually important ways and to make Book Three, not only a forum to offer new ideas, but acceptably complete in itself.

Since Book One contains many foundational beliefs, I've included the foreword to provide you with an introduction to it.

May your Faith In Life and devotion to it bless you with intelligent fulfillment, abundant joy, and placid gratitude for having been able to enjoy the gift of life.

FOREWORD TO BOOK ONE

The purpose of this book is to provide intellectually rigorous people with a credible, modern religion – a religion based on the appreciation, care, and enhancement of life itself, while it embraces individual freedom as a precious part of the natural expression of life. Free of superstition, it seeks to be entirely in agreement with truth as we know it and regards that imposing value as more than beauty; it is sacred – and no truth more so than that on which we base our religion.

Let us begin, in a manner similar to Descartes' often-referenced Method, with what we know for certain: we have life. Then let us derive from that foundational reality all of our beliefs.

I suggest that the basic tenet of such a religion would be Faith In Life – a trust in its greatness and logic – and that it contains within it the religious expression of Faith Through Life, that through the natural care and fulfillment of life we may not only preserve it and find considerate joy in it but also express our truest reverence to its Ultimate Source.

Frightening in its prospect, we seem almost on the verge of another Dark Age, while even the most enlightened among us often find ourselves adrift between religions that have lost credibility and our own shoulder-shrugging accommodations to hopelessness. I believe, however, that we have the ability to solve our problems, otherwise human life would be illogical, and nature is, if anything, precise to distant decimals.

How can we turn back the night? We can find, within finitude, the light of a religion that is equal to the challenges and joys of life because it grows out of life itself.

Such a religion can satisfy our minds, nourish our spirits, and inspire us to be knowledgeable advocates of life. It can impel us to devote ourselves, with mutually considerate freedom, to its care, preservation, and enhancement. It can enliven us with the will and wisdom to rescue life from nuclear war and incendiary terrorism, the diminutions of pollution and overpopulation, and the depredations of ignorance. It can inspirit us to accomplish the transformations necessary to secure an enlightened daily life for ourselves and our children and a promising future for as long as the earth may support humanity.

Let us begin while we yet may. It is not, as you know, necessary to kill a tree entirely to find oneself sitting by while it continues to brown into a lifeless remnant.

I have addressed a wide range of subjects, but I believe the diversity and not infrequent daring of my inquiries is what the challenge requires. May your life be blessed with as much true comfort and inspiration as I find in the thoughts that follow.

They are presented as they came to me. I have only applied some editorial niceties and a logical ordering of the parts. They are now as inviting as I can make them to minds informed, discriminating, and open to a vision equal to rendering our promise and obviating our peril.

FAITH IN LIFE

Faith In Life is the expression of our understanding of the greatness of life and of all Creation and our decision to trust in its greatness, care for it, and fulfill its finest possibilities.

We also understand that through the care and intelligent fulfillment of life, we can most credibly express our appreciation for life and our reverence to whatever its Ultimate Origin may be.

(2)

When we have Faith In Life, we have the ethical and religious answers we need. We trust in the greatness of life, devote ourselves to the fulfillment of its finest possibilities, enjoy the natural mental, physical, and spiritual rewards that derive from living so, and leave the struggles attendant to questions to which we cannot know the answers to the overreaching many.

(3)

Since we can write equations to describe Creation's behavior to distant decimals and life has evolved in ways that suit its forms to their functions, we believe it is credible to infer that Creation is logical. So we believe that what we can know at our level of existence is all we need to know and all we may be meant to know. This understanding is one of many that inspires us to have Faith In Life.

(4)

Faith In Life has the potential to rescue life and nourish life-oriented science, art, business, and everyday conduct. It can save us from our indifference to life, whether secular or sacred, give our history a new beginning and an assured future, and, by its logical extension of Faith Through Life, satisfy our longings in regard to communion with the Ultimate Source of life.

Further, we do not need to know about the Ultimate Source of life to express our faith in and reverence for who or what the Source might be, anymore than the recipient of a charitable gift needs to know the identity of his or her benefactor to feel an appropriate degree of gratitude.

(5)

When we believe in things that are good for life, we are, I think, more likely to make choices that help us, other people, and all other creatures have a good life.

(6)

Faith In Life is not an overly self-centered belief but an appropriately balanced one. Our sensitivity to life encourages fellow feeling. We also know that our own fulfillment has individual and mutual components, so we understand that we can only truly fulfill our lives when we are ever conscious of the fulfillment, not only of our own lives, but also of the lives of others.

If our Faith In [illegible] [illegible]amity, its naturally miraculous foun[illegible]ns ever before us and helps restore our [illegible]ence in it and our devotion to it.

(8)

Faith In Life enjoins us to live in ways that are worthy of life and of all Creation – for our own lives, other people, all other life, and all that has initiated and sustains life and the universe.

DEVOTION TO LIFE

Creation is devoted to life. We see sufficient evidence in the ways it evolves life wherever the conditions allow and conducts the processes to sustain it. When we live with Devotion To Life, we give our lives the same purpose as Creation's – the care, fulfillment, and creation of life (the latter being achieved by us, except for our scientific efforts, through procreation).

(2)

What else is the Universe about but devotion to the fulfillment of the possibilities of matter and energy, with its highest achievement the evolution and support of animate life, particularly animate life that can observe the processes of Creation, appreciate them intelligently, and devote itself, as all other creatures do in their own ways, to a role that is consonant with Creation's.

(3)

Let us decide to live in ways that make our lives meritorious with regard to life and Creation as we have come to understand them.

(4)

Conduct inspired by Devotion To Life enables, as what we may call Devotion Through Life, our most direct and credible active reverence to Creation, which we take to be the Universe as we can know it and any Creator or Creators who may be outside of our experience of it. It enables Devotion Through Life in much the same way that Faith In Life enables Faith Through Life.

(5)

Let's turn to our concept of Open Pantheism, which allows us to define an intellectually credible God. We say that the Universe itself can be seen as the grand molecular structure of our Natural Supreme Being, while we remain open to the possibility of a God or Gods outside of our experience, whose existence and character we can only intuit from what we can know about the Universe.

In the first instance, our lives and all other life are part of the life of our natural God, and, in the second instance, our lives and all other life are representatives of such a God or Gods. We can also see that the Universe is devoted to the evolution and sustenance of life, and so we must infer that any God or Gods who created it are devoted to its evolution and sustenance.

So when we live with devotion to life, we can very credibly infer that we have the same purpose as God.

(6)

We shall likely never know *why* life and the universe exist, except to intuit that existence is superior to nonexistence, especially existence that tends toward the creation and sustenance of animate life. We can express our logical response to existence and its most precious manifestation, animate life, by deciding to live in agreement with what are evidently Creation's own decisions about existence and life. They exist because Creation values them. We should value them, too.

(7)

What has been called the universal will-to-live can more exactly be seen as the vibrant manifestation of Creation's Devotion To Life.

(8)

We confront a vast and revered mosaic of traditional beliefs that are neither self-evidently nor provably true. Furthermore, at their core these beliefs are based primarily on reverence for another life, instead of for this life. We believe it is urgent to commit to the self-evident truth that animate life exists and the logic that, since nothing else would exist for us if it weren't for life, it is our fundamental value and merits our primary reverence. Then we will be inspired to live in ways that are worthy of it. Since life evolved with intelligence and sensation, we can infer that we ought to live it intelligently and sensitively – that is, to appreciate it, accept it, and devote ourselves to its intelligent and mutually considerate fulfillment.

(9)

We have, I think, mistaken the freedom with which the gift of life is blessed for the indifference of Creation. We have also not appreciated that this freedom invites us to devote ourselves to life, not because we are compelled, but though our own independent understanding and dignified decision.

(10)

Obviously, our first purpose must be survival; our second, fulfillment. This recognition helps us understand that our earliest ancestors had to, and the impoverished of the world still must, primarily struggle for survival. Since the appreciation of life evokes fellow feeling, we are conscious of our role in, not only fulfilling our lives in the developed world, but also in helping the world's poor put their lives on the sort of foundation that enables them to go beyond the struggle for survival to the enjoyable fulfillment of their lives.

(11)

Let us live with Devotion To Life, because the inherent value of life merits it, the conduct enables us to preserve life and enjoy the most fulfilling lives we can have, the conduct is in agreement with the devotion of Creation, and it is the most logical and direct expression, as Devotion Through Life, of our gratitude to and reverence for life's Ultimate Origin.

LIFE ITSELF

There are many things we cannot know for certain, but that we are alive is not among them. Life itself is, within the realm we can truly know, our sole verity and the rock on which we can confidently build lives of mutually considerate fulfillment, joy, and reverence, along with peacefulness at its completion. Let us finally commit ourselves to the beautiful and inviting, but woefully underappreciated and depredated, benefactions of Creation on this life-graced globe.

(2)

Life is the darling child of the universe and our own dearest possession.

(3)

Life makes sense as a self-sufficient, freely evolved, improvable gift. Such a concept also intimates a loving Creator or Creators – present, absent, involved or not.

(4)

Life itself may be seen as the immediate initiator and continuing enabler of all we can experience.

(5)

Life itself exceeds our wisdom and continues to reveal its promise. So beliefs we invent at any stage of our understanding are eventually not equal to it. The only belief that is equal to life is belief in life itself, which allows its potential to unfold naturally. All other human conceptions, being unequal to it, finally limit and harm it.

(6)

Life itself provides the natural standard by which our beliefs, finite and infinite, can be judged. To decide to dismiss it for any other standard is to decide we are smarter than Creation.

(7)

Life is the greatest existent Creation could evolve in the natural universe; it may, in fact, be viewed as the ultimate purpose of Creation.

(8)

Find me something greater than life to believe in, and I will show you an errant work of the imagination.

(9)

Opt for life while you and all of us have the opportunity. The act will make you precious and sacred to life's survival and fulfillment. It will, in fact, make you consonant with Creation, where, everywhere, we see its vibrant urge to evolve and sustain life and render its possibilities.

(10)

Let us make our religion out of life itself – the greatest existent we can know.

A SCIENTIFICALLY CREDIBLE DEFINITION OF GOD

The great question for religion is not whether or not God exists but how we may define what we consider a credible God. We can do so by utilizing what we can know. This approach enables our concept of Open Pantheism.

We consider the universe to be the grand molecular structure of our Natural Supreme Being, which evolves life as part of its processes, while we remain open to the existence of a Creator or Creators who are outside of our experience and may or may not always remain outside of it. This Creator or Creator also may or may not be within the universe and may or may not be involved in our lives, other than having originally created the universe that has evolved our lives and the processes that continue to sustain them.

We do not pretend to know the answers to these questions but believe we can express our gratitude and reverence without knowing them. We live in finitude and, logically, can and may well be intended to find our answers within it. We also believe we can credibly express our gratitude and reverence to life itself, the universe as our Natural Supreme Being, and any

Creator or Creators who may be outside of our experience without knowing anything that we cannot.

When we refer to Creation, we intend it to encompass both aspects of Open Pantheism – our Natural Supreme Being and any Creator or Creators outside of our experience, as in, let us live in ways that are worthy of the gifts of Creation.

KNOWLEDGE OF GOD

Scientific and philosophical truth grows out of our efforts to understand the universe, which we consider our natural God. So these disciplines are our natural ways to understand the mind and activities of God, at least, as exhibited in natural Creation. While we may feel and intuit more, we have no way to verify our conclusions and must be content with their tentative character. We can rightly know whatever knowledge is consonant with our place in the universe; the determination of its extent is part of our potential.

(2)

We can see God as the competent and generous Creator of life when we understand that God has evidently provided for all creatures to be able to function on their own and find joy in life. We can also intuit that God is self-sufficient.

(3)

Since we believe the gift of life comes with competence, we infer that God is competent; that life has been gifted generously, as generous; and, as a result, that life is good, and therefore God is good.

(4)

The only way to establish a benevolent God, whether such a God is still involved in our daily lives or not, is to believe that life has been created to be good. If it has been created to be bad, then the Creator must be bad, at least in relation to us, and therefore unworthy of our worship. But by the beauty and capacity of life, we conclude that life was made to be generally good and happy, although, within natural probability, at times bad and tragic.

(5)

A life-affirming religion is naturally disposed to see God in a way that is worthy of life.

CREATION

The concept of Creation is meant to include all that we can know or intuit, including the universe as we can know it and a God credible to our intelligence.

(2)

Life is Creation's great and generous gift and each life is a part of Creation's, or God's own, life.

(3)

We can express gratitude and reverence toward our Natural Supreme Being and a Creator or Creators who may be outside of our experience of it without knowing whether such a Creator or Creators actually exist or are involved in our lives, other than by having created them and having set up the system that supports them. By such an Open communion with Creation, we may satisfy, in logical proportion, our need to respect life and to feel at one with and commune with all aspects of what we realize we have been created by and are a small but precious part of.

(4)

Many seek absolutes, but Creation manifests values of balance. For instance, make gravity an absolute, and all would collapse.

(5)

It is logically appropriate to define "God" with uncertainty, in fact, the only definition that is consonant with our understanding of what God might be – the Power or Powers that are the font of all we behold, within it and/or external to it, presently involved or not. Fortunately, the riddle need not be resolved; a credible and lovable concept of God may remain the Power or Powers responsible for Creation as we experience it.

(6)

When we appreciate the greatness of Creation, we are inspired to have faith in it and to devote ourselves to the way we are a Creation of it, that is, to our lives.

(7)

Creation seems to have decided to let us discover the worthiness of its gifts, regardless of the destructiveness that might occur until we do – to trust its creations and the capacities with which it has endowed them.

(8)

Our surest communion with Creation is through our experience and care of life, its animate embodiment and representative.

(9)

We are conscious of ourselves and of the rest of Creation, and we are part of Creation. So, to the extent that we are conscious of ourselves, Creation is conscious of us, and to the extent that we are conscious of the rest of Creation, Creation is conscious of itself. We enable the appreciation of our own lives and the appreciation of the rest of Creation. In terms of this understanding, the most personal way we can talk with God is to talk with ourselves, and the most personal way Creation can be appreciated it through us. The communion, natural or beyond what we can actually know, is a holy mutuality. Not only we, but all species, participate within their capacities in this glorious, life-based communion.

(10)

Creation is adorned with expressions of abundant joy – the golden sun, the blue sky, the green plants, colorful flowers, the emerald seas, and the star-graced sky. We should live in contented consonance with its joy, knowing, of course, that there will be, in nature's probability, gray and stormy days.

(11)

See the joyful colors of nature. They are signs that life should be joyful.

(12)

Nature is enraptured with life. It explores its possibilities with ceaseless and spectacular vibrancy.

(13)

Revering Creation is part of revering what created it.

(14)

God may credibly be described as the Ultimate Source of life. As to the source of God, we do not seem capable of understanding a Source beyond what we consider the Ultimate Source; in fact, when God is seen this way, the quandary may be a contradiction that is a carryover from other ways of thinking about God.

(15)

It is contradictory to assert that we love God while we disdain what God has made. We best love God by loving the most precious thing that has been made, which is animate life.

(16)

Creation has made us who we are, as we have evolved and can experience our place in it, and has been special toward us in our unique endowments, particularly our intelligence, but it has also been uniquely generous toward every species. When we observe our own capacities and those of other creatures, great and small, we can see Creation's benevolence and sometimes, I think, Creation's smile.

(17)

Speaking with easygoing inexactness, the Deity and I have an unspoken agreement: He or She doesn't talk to me, and I don't talk to him or her, but I assume, if we can communicate, it is though life, the Deity by having made life possible and I by appreciating, revering, and doing the best I can with it.

(18)

The universe and life on earth were obviously here long before we were and will likely continue long after we complete our lifespan on the earth and, if we travel there, on other planets. The understanding encourages us to be more modest in envisioning the eternal Creator anthropomorphically. He, She, It, or They existed billions of years prior to our arrival and will likely continue billions of years after our departure. But we do have a place in the whole, and it has been left to us to sanctify, enjoy, and distinguish it.

(19)

When we create something, what delights us? Seeing it perform well. So, to the extent our own perceptions can tell us, we can delight our Creator by performing well.

(20)

It seems logical that we can express love to the Creator of life by succeeding in life, and whether or not the

Creator is conscious of our love, beyond how we ourselves are conscious of it as individual embodiments of our Creator, is essentially irrelevant. We can express it based on what we know we have been gifted with, and we acknowledge those gifts as part of our abiding appreciation of Creation overall.

(21)

The natural arrangement of Creation appears to indicate that each type of life has evolved to enjoy life in its place as a complete gift.

(22)

All animate life has been evolved, at least, immediately, by our Natural Supreme Being, the Universe, and is part of our Natural Supreme Being.

(23)

Everything moves because everything is alive – from subatomic particles to galaxies.

(24)

We are created in the image of God in the sense that we are part of the natural universe, which we consider our Natural God, and therefore we are part of the image of our Natural God. If there is a God or are Gods beyond our Natural God, then we can be seen as expressions of Divine intent.

(25)

We personify God as a reflection of our ideals; for instance, in tribal times, as a God of power; later, of love; now, as the Origin of life.

(26)

Our role in Creation is to adjust appropriately to our place in the great unfolding of the life in the universe and to live our own lives in consonance with it, while we express gratitude for life and the unique gifts we have been endowed with and allowed to enjoy during our opportunity to live.

(27)

When we appreciate life, expressing our gratitude to Creation seems a natural aspect of our lives.

THE NATURALLY UNFOLDING UNIVERSE

The way the universe is unfolding seems to be consonant with our intuition that it was designed magnificently and then allowed to evolve in freedom, as if God opened his hand and let it spring to life.

(2)

The processes of the universe seem to be based on life-evolving, life-sustaining, and life-completing behavior, with the fundamental principles inherent in matter and energy and manifested in subatomic particles, atoms, and molecules; then in cells, organs, and individual animate life forms.

(3)

If the Big Bang and the expansion of the universe are conception and growth on a cosmic scale, then the entire history of the universe might well be, overall, as simple as the opening and the closing of a hand – a new universe with a lifespan like everything in it.

If this is the case and if a perfection of a perfect being is eternity in time, then our universe may restage a Big Bang and expand again, as part of an endless cycle, or be part of a larger whole that creates or renews its parts in an infinity of time.

(4)

While there are apparent instances of imprecision or unpredictability in the behavior of subatomic particles, or what we perceive as imprecision and unpredictability, the universe is, at the level we experience it, astonishingly precise and, to the extent we understand its processes, naturally predictable. It displays, to the extent we can perceive it, a majestic orderliness in its arrangement and processes – in fact, enough to nurse up, nourish, sustain, and, in time, conclude planetary life. We may say, in fact, that its precision is part of what makes life possible. If gravity, for example, varied much, the atmosphere might unpredictably float off the earth. And, if our biological processes varied much, we might never experience equilibrium, let alone the sense of physical wellbeing that healthy people generally enjoy.

(5)

When we see the great urge to animate life on the earth and known the relative uniformity of the sun-system universe, we can more confidently intuit that, among the billions of stars, around which more billions of planets have formed, the same urge toward animate life exists. The perception that all the potential for it would remain inert contradicts our knowledge that, in terms of what we call matter and energy, the entire universe is alive and seems intent on evolving animate life wherever possible, suited to the conditions that obtain. We may, in fact, see ourselves as the latter-day equivalents of the populace of the old world, wondering if there might be something beyond the far horizon, before the new world was discovered. One day, I think, we will be advanced enough to join, through electronic communication and later perhaps in person, the community of the universe.

(6)

While the complexity and beauty of the universe are traditionally invoked to prove the existence of an intelligent designer, I prefer to invoke the wondrous attributes of the universe to persuade us that it ought to be reverenced primarily for itself and then, secondarily, for the Ultimate Source of the intelligence manifested in it.

(7)

We are not quite satisfied with the design of the universe and the animate life it has evolved, yet even the design of a subatomic particle is beyond our understanding.

(8)

While the progress of knowledge and insight has yanked the tent pole out of Thomas Aquinas' five proofs for the existence of God, it has also provided us with a new and more credible interpretation of "the proof from design."

By revealing the wonders and majesty of natural Creation, science has furnished us with "proof" that natural Creation of itself merits our faith, devotion, and reverence, along with providing us with "proof" that any Creator or Creators who may be beyond natural Creation also merit our faith, devotion, and reverence.

The first, based on respect for the natural world, will likely find more agreement among scientists than the second and, in practical terms regarding the preservation and fulfillment of life itself, is more essential, while the second is an intuition that will likely be welcomed by those who find it consonant with reason and their hopes. As expressed, both are consonant with mine.

(9)

Three modern kinds of reverence are the scientific study of life and the universe, which is also the study of the true ways of the Immediate and, if it exists, the Ultimate Source of life; the artistic renderings inspired by life and the universe, which constitute praise to them and to their Source; and the life-devoted conduct of everyday life, which is the lifelong reverence all of us can express to life and its Immediate and Ultimate Source.

(10)

When we appreciate the enormous cooperative effort that is required – by atoms and the solar system, by our galaxy, in which the solar system finds support, and, in fact, by the entire universe – to create and sustain even one animate life, we realize the extent to which the universe is devoted to the evolution, sustenance, timely conclusion of, and the renewal of animate life.

(11)

Intelligent beings on other planets no doubt have imagined, in their intellectual infancy, their own divine beings, and it will be amusing to see how much they are similar to or different from the ones we have imagined and which planets and at what stage have converted to belief in life itself.

(12)

While the universe is billions of light years in size, many people still spend more time wondering, not about the wonders within it, but about existents that may or may not be beyond it.

(13)

We should see the order, intricacy, and beauty of the universe and the miraculous capabilities of its creatures as evidence of the value of life itself and the extent to which Creation values it. Then we will be inspired to understand that life itself has more than enough value to evoke our reverence, convince us that it is a sufficient possession in itself, and understand that revering it is the most direct communion we can experience with its Ultimate Creator or Creators.

THE UNIVERSE AS OUR NATURAL SUPREME BEING

The Universe, as our Natural Supreme Being, may be seen as either God or the supreme representative of a God or Gods who are not part of our experience of it.

(2)

Some may say, why can't we simply decide there is no God? The decision is illogical and ultimately unsatisfactory. We ought rather to define what we mean by "God."

If we're speaking of a "supernatural" God, the decision about such an existent requires information that we do not have, will likely never have, and, we must suppose, are not required to have. We must also ask what we mean by the supernatural, an idea that, by definition, demeans the natural and, as far as we can know, may not even exist. We can only be open to whatever the truth might be.

We ought also to understand that the Universe itself is certainly magnificent enough to accept as our natural God.

So, as to the existence of God, the only speculative aspect is whether the Universe is the self-created structure of our one true God, of which our lives are each an expression of its own life, or the representative of a God or Gods who are outside of our experience.

(3)

When we understand the greatness of the Universe, we realize its capacities are far more astonishing than the feats of the Gods we have invented. The Universe does not just appear in a burning bush and carve a message in rocks. It evolved and is the fire and the rocks. It does not just part a river. It created and is the river.

(4)

The Universe always behaves naturally, which manifests a greater wisdom than is apparent in the storied activities of the Gods we have invented. Its more predictable behavior also has effects that are more benevolent.

(5)

The Universe – or the entity of which it may be a part, the Multiverse or Megaverse – is likely a closed, if growing, system of matter and energy, which we consider its fundamental substance and life, functioning internally in a perfectly natural way, at least to the extent we can perceive its workings on our level. This natural self-sufficiency is one of its most blessed attributes. It provides a trustable stage for the animate life within it.

(6)

Rightly appreciated, our Natural Supreme Being is greater in its natural achievements than any supernatural power humans have imagined. By appreciating its natural powers, we can infer that if anything has what we consider supernatural powers, it does or its Source does. Compared to creating a galaxy, how miraculous is walking on water?

(7)

Life is either an attribute of the Universe, as our natural God, or an attribute of it and a living representative of a Source or Sources outside of the Universe. Either way, for practical ethics, life is God and, similarly, God is life.

(8)

Everything we behold and experience is part of the life and substance of the Universe, which we consider our Natural Supreme Being.

(9)

The universe is, in a real sense, much like each of us – primarily matter with a spirit of life. Its life derives from, or antecedes, the birth of the universe, while each of our lives is descended from, animatedly, the first spark of animate life on the earth, which, in turn, may be traced, inanimately, as far back as the life of the universe.

(10)

In terms of our Natural Supreme Being, we are not estranged from God but part of God.

(11)

When, as a living part of our Natural Supreme Being, we talk with ourselves, we are, in a very fundamental sense, talking with God.

(12)

A starry night lets us see more of our natural God. It is a partial view of a vast structure that has evolved and supports our life on earth. It is our complete, unmagnified view of our natural God, while the earth itself represents our immediate view.

(13)

When we do not appreciate the greatness and sanctity of life, we search for a conception of God and arrive at such relatively vacuous concepts. Take Tillich again and his concepts of "the God beyond the god of theism" or "anything you can believe in." Such notions can hardly satisfy the longings of most people who are seeking spiritual nourishment. They also help us see how blind even the theologically knowing have been to the natural Godliness of life itself.

(14)

I would think that the life of each planet is as unique as the life of each person but, as people are similar in more ways than they are different, I assume that the same pattern applies to planets. This vision means that we have evolved as the appropriate inhabitants of our sun-earth relationship and that similar sun-planet relationships would likely evolve forms of life similar to ours.

(15)

The Universe manifests ongoing fulfillment and potential, which constitute the past, present and future achievements of its lifespan.

(16)

The Universe is evidently devoted to the fulfillment of its possibilities, and the human race, as part of it, should join in.

(17)

The belief that we are at one with our Natural God is redemptive. It is a basic and abiding satisfaction, while belief in only a supernatural God, who may or may not be looking in on us, seems, by comparison, secondary.

(18)

The conception of a Creator of the Universe, other than the Universe as its Self-Creator, posits an awfully powerful invisible or external existent, but the Universe as the maker of itself, while enormous, brings us to the immediate creators of its life, which are invisibly small: subatomic particles, the atoms they form, aggregates of atoms into molecules, and molecules into cells – matter and energy, or substance and life, guided by the inherent intelligence of the Universe, self-evolved or placed within it.

(19)

As we believe body, spirit, and mind are different aspects of one existent, namely physical existence or the body, at least, in natural terms, so we may infer that the matter, energy, and intelligence of the Universe are aspects of one existent, at least, in natural terms.

(20)

As the beauty and talents of life forms often exquisitely indicate, even among the wiliest, they are perfections of our Natural Supreme Being or of it and a Creator or Creators beyond it.

(21)

A star is like a supreme nucleus in the grand atomic structure of a solar system.

(22)

Having detected the microwave remnants of the Big Bang at the edge of the Universe we can know, we have confirmed that the universe is, as a naturally functioning system, finite. Recent research seems to indicate that it may also be part of a larger system, which we suspect, while growing, is also finite.

We might ask, does being finite make it less? I don't think so. While Spinoza decided that infinite extension is a necessary attribute of a perfect being, we might just as well say, in the ragtag of ill-defined concepts, that the power of self-definition, which includes the power of self-limitation, is also a necessary aspect of a perfect being.

(23)

The knowledge that all in the Universe is part of God or the representative of God is actually the most personal concept of God, because now God or our closest relationship with God is our very selves and all other life, even the insensate stones that provide solidity to the earth, our basic home in the Universe.

(24)

Glancing starward through a cloudless night and noting the enormity of Universe, we sometimes feel so glad that we were born that we're inclined to bow or genuflect, but most likely we merely sigh with gratitude, express our awe with a reverential smile, and then move on.

A CREDIBLE COMMUNION

We can participate in a credible communion with and through what has been created. Such a communion has two aspects – immediate and ultimate. Our immediate communion is with ourselves, other people, all other creatures, and with the universe as our Natural Supreme Being. Our ultimate communion is through all that has been created to our Natural Supreme Being and to any Creator or Creators who may be outside of our experience, except as they may manifest their presence in what has been created.

We can also indulge in, if we're inspired to, a less credible communion, which may seem to be more direct but actually isn't; for instance, we may feel like waving at the clouds and saying thank you. But we also ought to keep in mind that any God or Gods we hope to commune with, other than our Natural Supreme Being, the Universe, may not be peeking out from behind a cloud.

OUR EARLY RELIGIONS

Before we begin to express our thoughts about our early religions, let us note that the religious sentiment is not identical with the different religious beliefs we have developed. The sentiment is an understandable human response, based on our place in existence. We find ourselves with life in a world we have only recently begun to understand. Traditional religions are relatively early expressions of our response to existence.

Our ancestors saw that they were the benefactors and victims of many things but did not understand why. Among them were health and disease, plenty and famine, safety and accidents, birth and mortality. Since they did not yet know how to investigate real causes, they accounted for these "blessings and curses" with reasons they imagined.

Today, we understand the causes for many of the things we are the benefactors and victims of. Since we have been able to alleviate many of the liabilities, we have been able to enjoy the benefactions more.

We have also learned that we live within natural probability and understand that it is bound to exist in a setup that enables the independent development of animate life.

We have come, in addition, to understand the concept of lifespans; that each life consists of conception, development, birth, growth, fulfillment, and completion. We realize that each life has a potential for fulfillment and that the usual healthy lifespan of each species seems to allow for the realization of it. We appreciate, too, that the concept of lifespans enables a place for new life. While we may at times regret that mortality is not otherwise, we must finally recognize that it is a wise aspect of life. What is our appropriate adjustment? We may diminish its effect and thereby more tranquilly accept its inevitability by fulfilling the potential of our lives.

Unlike our ancestors, we no longer need see mortality as an unacceptable liability against life. We can admit ample merit in life and rescue it from the debilitating grip of mortality. We can decide to live for this life – fulfilling our individual and mutual potentials as best we can.

Do we have to give up on the hope of any life there might be after this life? No, because all we can know pertains to this life. Whatever is beyond our understanding will always remain beyond it. What we can understand is that this life is the one we can truly know, care for, fulfill, and enjoy. We can also appreciate that, should there be another life after this one, the logical way to merit it, if it can be merited, is to take good care of this life first.

So we come to a new question. Can we develop a religion for life as we have come to understand it? Briefly, yes. We can have Faith In Life, as an expression of our trust in its greatness and our reverence to its Ultimate Source.

(2)

I prefer to emphasize a new way to believe, because I know that is what we need and because the arguments against our early religions have been made. But to place a life-oriented religion on a firm foundation, some recapping is necessary. I trust those who disagree will understand that my thoughts are meant only for those who are seeking a new and intellectually acceptable home for their religious sentiment.

(3)

As soon as we say our beliefs are unchangeable, we limit Creation to our understanding at the time. We should ever live in wonder of it, learn what we may about it, and believe in ways that are consonant with our evolving understanding. Such openness to life is the only way for us to have beliefs that are equal to life and its ever-sacred capacities.

(4)

The inevitable threats to the Roman Empire, in a world widely barbaric, and ultimately the destruction of it, precipitated a widespread retreat from life and led to the sullen otherworldliness of The Dark Ages. It is time, in fact, urgent, for humanity to recover from the trauma and turn back toward life, before our relative indifference to it enables its widespread devastation and perhaps its annihilation.

(5)

Tolerance, freedom, and love invite us to believe that we each have the right to believe whatever we choose, as long as our beliefs do not hurt others. In fact, true sympathy advises us to bless whatever beliefs help any of us navigate life's ups and downs, providing that the beliefs are not inimical to those who believe differently.

(6)

The established faiths have existed for many centuries. We have now had a long period in which to measure their effect and must admit that they have not been able to transform human behavior to an acceptable degree. We are still outrageously, and perhaps ruinously, destructive of life.

These faiths have revealed, whatever we may find in them that is positive, that they do not have the capacity to remediate destructive human behavior, let alone to convince the majority to sanctify life and devote themselves to it. Although still presented as our salvation, they evidently are not credible or persuasive enough to rescue us, particularly from ourselves.

Since we now find ourselves on the brink of such calamities as nuclear war, environmental collapse, or a murderous reversion to divisive barbarism, the development of a new religion, or philosophy, that may prove equal to our challenges has become, not an option, but an emergency.

(7)

Rituals based on incredible beliefs are often more concerned with propping up the beliefs than addressing the needs of those in attendance. This shortfall is never more apparent than at the most emotionally consequential occasions, such as a funeral service. The intent to reinforce a fragile vision tends to make incidental even the person the occasion is meant to eulogize.

This frequent practice creates, to my perception, a callous undercurrent that says the foundational beliefs of the faith are far more important than the necessarily transient members of the congregation. How much more inspiring and nourishing a service would be that is primarily about the person or the people who are the subject and auditors of it.

(8)

Could a God equal to Creation require or be pleased by the sycophantic adoration and protestations of guilt expressed in traditional religious ceremonies, when even a normally competent and generally well-behaved individual may be made uneasy by the pronouncements?

Have we not evolved to be more than slavish importuners, egregious flatterers and helpless wrongdoers before whatever Deity we hope to please? Are there not more distinguished and wholesome ways to converse, if you will, with Creation?

(9)

Does it make sense to praise the Creator at one moment and in the next to demean what has been created? To laud God as the creator of life and then to wail that life is unsatisfactory? Would a good God provide us with inherently evil natures? What would such a God prefer intelligent creatures to feel – debilitating guilt or temperate pride?

(10)

While most consider religion to consist of the sanctification of inalterable traditions, a religion equal to our potential would alter to accommodate our growing knowledge.

(11)

If we are driving toward a destination and learn that our directions are faulty or that it may not exist, would we not seek other directions and alter our route?

(12)

Superstitious beliefs are uncertain and deny the adequacy of naturally credible beliefs. Most regrettably, they substitute for a true appreciation of life and block participation in it.

(13)

Any belief that assigns greater value to life after death than to life itself subordinates life and predisposes it to mistreatment.

(14)

It has been said many times that the belief all humanity has been exiled here as a place of atonement for the apple-size sin of Adam and Eve posits a God who, incongruously, knows nothing of mercy and moderation. It also fails to account for why all the other forms of life have been cast out of Paradise with us.

We may, thinking better of ourselves and of any God equal to creating us, reject such a life-demeaning tale and allow the company of the multitude of other innocents to reinforce our belief that animate life has evolved here, not as a punishment, but as part of the sacred life of the universe. We are more worthily here to experience, fulfill and enjoy our share of the gift of life, in harmony with the unfolding of life in the universe.

As often noted by others, we also ought to suppose that an all-powerful and all-good God could imagine a more benevolent – and considerate for children of all time – way to redeem us than to decree the torture and death of his own son, even if his death was only intended to last for three days.

Let us think more of life and Creation, so we may rescue them from all such diminutions.

(15)

What are we to think of such outrages as the injunction to slay every man, woman and child; the flames of the Inquisition, and the current Muslim disgrace that the Deity rewards suicide and mass murder? All of them are instances of irreverence for life. If God can weep, such sins against life would certainly provoke copious tears.

(16)

Is it truly blessed to dismiss the natural universe, along with the wonder of all of its life, and regard what might exist outside of our experience of it as more worthy of our devotion? Isn't such a life-dismissive view based on insensitivity to the greatness of life as it has been gifted and to the Ultimate Source of life? Is such a view likely to invite the commendation of Creation?

(17)

It would be a religious act for all monotheistic religions to agree that they all, of necessity, worship the same God.

(18)

What should be the primary calling of religion – to enable reverence for all that has been created or to present itself as a remedy for death? Should it laud the blessings of life or seek to compete with them? Should the many good people who could help rescue this life convert to belief in it or continue to devote themselves to reverence for another life?

(19)

I do not often quote other theologians. We are usually on extraordinarily different pages. But let me mention Paul Tillich, who refines belief into the thin milk of telling us that religious concerns must be "ultimate" and exclude "all other concerns from ultimate significance." What have we here but the absolute rejection of the significance of this life? If we do not go beyond such benumbed abstractions, our survival, let alone the fulfillment of our finest potential, doesn't have a prayer.

(20)

Our ultimate concern should not be for a conjectured exterior God and an uncertain afterlife but for the world we have been born to and the life we find Creation has placed in our trust.

(21)

The humanness of us all becomes clear when we realize that that none of the "Gods" who appeared to our credulous or calculating ancestors knew more about the natural world than was known at the time. For instance, none mentioned the possibility of antibiotics and microchips.

(22)

We have sympathy for the plight of our ancestors. Obviously, one reason that they demeaned life is that they were able to reduce the impact of its loss, at a time when their lives were far more insecure and brief. Their adjustment is altogether understandable. But for us to continue slighting life when it has revealed, for example, that we can cure many diseases which were formerly fatal, is an expression of irreverence for life and, given how desperately life now needs us to commit to it, a twofold irreverence.

(23)

Many who would uphold our traditional religions exert themselves to demonstrate that there is a God, and some philosophers allow the discussion of religion to revolve around the subject. But whether God exists in the traditional sense is not actually the fundamental question. We can define what we mean by a credible God; that is, the Ultimate Source of animate life and all else that we behold, whether it is what we call the natural Universe or a God or Gods outside of our experience. But such a concept of God is hardly coincident with the God described by the books of the religions we have inherited. They are, in fact, two very different considerations.

(24)

There can be no argument about the existence of an Ultimate Source of life, because we define it so that it merely refers to whatever initiated the universe and then its animate life. We may, in fact, even one day discover, through science, that there is a natural Ultimate Source of life. Yet we are likely never to know anything for certain about a God or Gods who may be outside of our experience.

(25)

Most people who have faith in our traditional religions are among the nicest and most ethical people in our society and potentially among the greatest advocates life could have, if only they could see that putting life first is what life itself merits and what the Creator of life would most likely want us to do and reward us for doing, especially now, when life is endangered.

(26)

Since our traditional religions have failed to establish the sort of world all responsible people long for, what may we conclude but that they are inadequate to the task? A telling question presents itself: how can any religion enable a world that is equal to its finest potential when its primary concern is another world? The primary concern of the religion we need is, obviously, the world we inhabit.

A MODERN RELIGION

While traditional religions are based primarily on what is beyond the natural universe, Faith In Life invites us to make our primary concern what is within it – in fact, within a tiny portion of it, our blessed little isle of animate life in the uncertain cosmos.

(2)

We search for our religious beliefs beyond the world we can truly know when we have only begun to understand the wonders within it, which are very natural miracles.

(3)

A religion based on the appreciation of life, not the inevitability of death, is far more nourishing, individually and socially, and provides a healthy guide for our daily conduct and a credible foundation for us to feel at one with all of Creation.

(4)

We tend to think of being religious in terms of adherence to one of the religions we have inherited, but the religious sentiment exists in its own right. It is an expression of our natures in the situation in which we find ourselves and merits a religion that is worthy of it – one that is equal to our growing knowledge and that satisfies our longing for communion with all we can perceive, as well as intuit.

Briefly, what do we know? We find ourselves with life, in a universe we know is majestic, while a power beyond our experience does not seem to influence our lives, at least, beyond whatever natural role it may play in evolving and sustaining our lives.

Given the limits of what we can know, it is easy to understand a contemplative person's inclination to Deism or agnosticism. Atheism, on the other hand, seems to me to be the alternate of fundamentalism; both extremes require definite knowledge that is simply not available.

Is it possible to allow that we have never really heard from what we consider a supernatural world and yet describe a complete religion? Of course. In fact, I think the very silence indicates that we are meant to find all of our beliefs within the natural world. They would be based on life as we can know it, so they would enable us to be intelligently religious.

We can, in fact, go beyond our very natural uncertainties and establish a religion founded on the worthiness of what we can know, within life and the universe, while we also allow for an expression of reverence to whatever the Ultimate Source of what we experience may be.

It is not necessary for us to know for certain whether the Ultimate Source of life is the Universe itself or a God or Gods outside of our experience. We do know that the immediate Source of life is the natural processes of the Universe. We cannot and, we must conclude, need not know more.

(5)

We have a natural, healthy desire to feel at one with Creation and, to the extent we can, communicate with it. Why wouldn't we? We're part of it, that is, we evolve out of it, live as part of it, and will return to it. But is there a credible way to satisfy these longings? Yes, but not by looking outward, beyond the situation in which we find ourselves, but by looking inward, toward our situation and through it.

(6)

We are alive and within the natural universe. So we can have a credible natural feeling of oneness with life, as well as with the universe, which is what, in the world we can know, enables and supports life. But first we must desire to have a communion with life and the universe.

How can we enable it? We must get past our traditional negation of their value by appreciating them as we have come to understand them. When we do, we can respect them; in fact, we can revere them. What are we to revere? The gifts of life, such as our ability to be cognate and sensate, among other beings who also have a share of these gifts, within an environment that has evolved us, nourishes us, and continues to support us.

Appreciating these gifts, we may revere them. Viewing the universe and knowing it has evolved us naturally, we can revere it. Since we live within life and the universe, we logically express our primary reverence to them.

Yet we wonder if there may be a power or powers beyond our experience that have ultimately enabled life and the universe. We may also revere it or them, and to do so we do not need to know anything certain about it or them. We can do what our situation invites us to; that is, to accept the ultimate mystery of existence. When we do, everything we believe can be credible, and we can experience a complete and credible communion with what we can know and with what we can only intuit the existence or nonexistence of.

(7)

When we realize that life and the universe are the greatest existents we can truly know, we can naturally desire to be at one with them. We can also learn to have great faith in them – a wise trust in their greatness – and realize that through life itself, which is what has been created, we can most credibly express our worship, not only to life itself, based on its own merits, but to the universe, as its immediate and knowable Creator, and to any power or powers beyond our certain knowledge that may be the ultimate Creator or Creators of life and the universe.

(8)

The contemporary ethicist, Peter Singer, notes in discussing when euthanasia might be ethically justifiable, "We usually value life because it is the basis for everything else that we value, whether it be happiness, appreciation of beauty, creativity, love, or the exercise of our rational faculties." He does not explore the implications of his realization but goes on with his main subject. Yet he does note the missing concept in philosophy. It is gratifying to see that others are realizing the deficiency. I've often felt alone in my appreciation of life since I realized its ultimate value, which was when I was in my late twenties. Expressing the beliefs that spring from the realization has been a long and generally individual devotion.

(9)

Of all the rooms in the mansion of human knowledge, religion is the one with the most dust. The fact that we have allowed so essential an area of our intellectual capacity and spiritual distinction to molder remains a growing incrimination of our willingness to be ethically responsible for religion. Although tradition informs us that religion is an inalterable legacy, by now we, or a great many of us, can see that religion as we have inherited it is afflicted with gaping challenges to the minds we have evolved with and to whose capacities we owe a sacred allegiance.

We must conclude that the room in which religion abides ought to be kept up as much as any other room. If religion is to be equal to our capacities, it should evolve along with our growing knowledge and more acute thinking. Only then can its contents be consonant with our life-given capacities and our best efforts on behalf of life.

Let us finally throw open the doors, let in the light, polish up what is there and, most urgently of all, bring in new belongings, so we may furnish it with an authentic and intellectually dignified spirituality – a religion based on truth as we have come to understand it. It is the way we can give ourselves a religion that is worthy of our religious sentiment.

Faith in our best efforts to do so is part of our trust in Creation and the intelligence we have been endowed with.

(10)

The hopes of religion are timeless but the tenets improvable.

(11)

I am ever astonished that natural Creation itself does not seem sufficient to most people as the foundation of religion.

(12)

Humans have knocked on locked heaven's door for all of our written history. While there have been many reports of an answer, none has been universally recognized, because all of them contain notions that falsify them. Moreover, none of the hopes or admonishments has been able to reform humanity sufficiently to assure even its survival. Let us desist so that we may discover the answer that has always been before our eyes.

(13)

We know that the universe, as we experience it, is so logical we can write equations to describe its functions. Yet we do not know and may never know all the capacities of it, which may or may not transcend the most treasured hopes of traditional religions.

(14)

Traditional religions do have tenets that can contribute, in natural ways, to a life-gracing religion, such as "You shall not kill," "Love your neighbor," and "Blessed are the peacemakers." They and those who espoused them can become part of our common human heritage.

(15)

A life-devoted society would, I think, find delight and comfort in such salutations as "Life bless you" and "Life bless you and keep you."

(16)

An appreciation of life's greatness establishes a sound and happy relationship with life.

(17)

Enough pain and loss can drive nearly anyone to make a plea to God for mercy, whatever his or her idea of God may be. Yet our truest communication with God is very likely not beyond life but through it, not by appealing to an uncertain hope out there somewhere but by looking in at life and its abiding gifts. In Faith In Life and its extension, Faith Through Life, we may find, even in pain and loss, our most credible comfort.

(18)

When people search for God, they deserve more than the hope of finding one we've imagined. They deserve to know that they can commune with whatever made life by appreciating life and, if they wish to please whatever Ultimate Maker life has, by reviewing their behavior and vowing to live in ways that are worthy of Creation's natural gifts.

(19)

What about our children and the religion we teach them? Don't they deserve to be taught a credible appreciation of life and made aware of the natural invitation they have to do their best with it? Otherwise, they often learn to be skeptical and lack direction. They can become, instead of found and happy souls, lost and sad ones – and in time adults, not devoted to the care of life and individual and mutual fulfillment, but careless of life and devoted to self-interest and, at worst, outright destructiveness. We owe a life-enhancing choice to our care of them and to our own achievement.

(20)

Faith in life itself provides the natural foundation for an intellectually inviting, scientifically credible religion.

(21)

What are greater forms of worship – the acceptance of life as a gift, the devoted care of it, gratitude for human capacity, and confident life-gracing achievement or the stigmatization of this life as a sinful punishment, relative disregard of it, guilt over human endowments, and protestations of helplessness?

(22)

Our infinite concerns are best addressed though finite means.

(23)

While traditional religions primarily point us out of this life for our ultimate answers, Faith In Life primarily points us toward this life for them.

(24)

Even the shortest life is sacred; so is even the smallest thing that supports life.

(25)

What is a holier activity – the considerate fulfillment of this life or devoted prayer for another life?

(26)

While reading books, reciting prayers, singing hymns, and performing rituals may be part of religion, our most basic and urgent reverence is expressed by living in the service of life.

(27)

All credible reverence is toward life and through life. This truth is redemptive, not only intellectually, but spiritually. It is also redemptive in a very practical way, because it inspires us to do our best in, with, and for life.

(28)

A religion founded on life itself is finite at the same time it is infinite. It is directed, first, at life and, second, through life to its Ultimate Origin.

(29)

To know the extent to which life is a blessing is to know all we must and are meant to know. Imagine a life-devoted world, and the validity of the statement becomes immediately evident – humans worthy of the sun, the verdant earth, the blue sky, and all the forms of life.

(30)

Let us imagine a world of good works. Then we can see clearly that we most effectively express our reverence through works.

(31)

Each of us is an embodiment of Creation's own life and a living representative of the Ultimate Source of life. We each have life-endowed capacities that allow us to actuate the finest qualities inherent in Creation within the locus of our existence. Doing so is our truest worship of life and its Origin. When we live in a way that is completely informed by these truths, we may, I think, be considered saintly, more exactly, saints of life and of all Creation.

(32)

It seems self-evident that whatever created life would most admire those who devote themselves to it and find satisfaction in its care and fulfillment.

(33)

The true voice of Creation is made up of the voice of every person, every other animal, even the leaves, the breezes, and the waves. It is the complete orchestration of Creation.

(34)

The perennial good of religion is that it generally attracts people who are well intentioned. We are all too aware of the exceptions.

(35)

By caring for life and fulfilling it considerately, we may each embody a life of natural worship.

(36)

Whatever has allowed us to evolve this long without unnatural influence is, I think, unlikely ever to attempt it. The time has already been long and the provocations numerous.

(37)

Faith In Life is, I think, the most direct way to show reverence for what made life.

(38)

True holiness consists in living in harmony with life, not rejecting it.

(39)

Let us imagine a place of worship for life itself.

Visually, it would present the beautiful aspects and events of life and might be graced with some plants. Any altar or tabernacle would feature the basic symbol of life. (So traditionally unmindful of life are we that we do not yet have a universally recognized and adequately appreciated one. Where is the artist who will create the appropriate and worthy symbol of life?)

The ceremonies would, of course, be devoted to life. They would be centered around such topics as the natural wonders of life and why life invites our faith in it and devotion to it; the transcendent value of each life and the natural rewards of fulfilling its finest possibilities; the ways mutual consideration and community life are a necessary part of each person's fulfillment; how all the forms of life invite us, through their wondrous capacities and importance to our own lives, to reverence them; celebrations to mark the great events of life, including birth, maturity, marriage, and life's completion; and a clear understanding that through Faith In Life we can best be inspired to care for life, find joy and satisfaction in it, show our reverence toward its Origin, whether the natural universe itself or the natural universe and a Creator or Creators beyond our experience, and know, since we have done the best we can with life, we will have, logically, merited, if it can be merited, any afterlife there may be.

What would we call such a place of appreciation, guidance, and worship – a temple of life, a church of life, a congregation of life, or simply a place for life – would be incidental. Far more important is that it would be a spiritual home for life and its intelligent appreciation – an environment devoted to and sacred to life.

Astonishingly, it would be the first holy home for life that the human race has ever known.

(40)

As I have said similarly elsewhere, in appreciating the natural world, we may consider the universe to be our natural Creator, the earth lit by the diverging rays of the sun our natural temple, the congregation all the forms of life, and the clergy those who understand the blessings of life enough to represent them credibly and invitingly.

Of course, it would be commendable and socially inviting to attend religious services intended to celebrate our appreciation for life, and attendance would have appeal to many. Yet it is also worshipful just to wake up, for instance, on Sunday morning, and take a few moments to appreciate life and perhaps ponder with grateful wonder what created it.

(41)

When we live in the service of life, we participate in an ongoing religious service.

(42)

Who would a missionary of life be? A person devoted to enlightening others about the appreciation of life and the sanctity of having faith in it. He or she would advocate the care, fulfillment, and natural completeness of life and thereby inspire others to commit themselves to life, so that they can also enjoy mental achievement, physical normalcy and joy, and spiritual health. Such a missionary would encourage and laud every form of mental distinction, champion the natural enjoyment of sex and other forms of life-enhancing physical activity, true communion with ourselves, with one another, and with all of life; note the individual and social benefits of mutuality and convivial conduct, and advocate life-sustaining practices and environments. The missionary would suggest, not command, because he or she would understand that freedom is necessary for the realization of human potential. He would be, at his best, a saint of life to himself, to all other life, and to all of Creation.

(43)

Who is more appropriately a spokesperson for religion – a person who rejects the gifts of Creation as inadequate and espouses the hope for more or a person who accepts them as satisfying and espouses the care and fulfillment of them?

(44)

I believe the holiest people are those who express their faith through the care and fulfillment of this life.

(45)

Life itself is a naturally universal religion, for the simple reason that we are all alive, but our knowledge that life can only exist as it has been gifted, in freedom, requires that we acknowledge freedom of religion for all.

(46)

We show our basic reverence to Creation by the appreciation and care of its gifts.

(47)

Life-oriented beliefs are harmonious with Creation's great will to live, evident in the sprouting of life wherever conditions allow, for instance, of a plant in what seems an impossibly arid environment. Such beliefs enable the human component of Creation's overall will to live. Death-oriented beliefs are contradictory to Creation's great will to live. Such beliefs hamper the human component of it.

(48)

Consider how enormous and filled with wonder we have learned the natural universe, or the megaverse of which it may be part, to be. Might it have the power to generate itself and to create an appropriate place for us? What if it is the creation of something outside of it? How powerful, and likely enormous, would such a power be? Is there a scale on which we might come face to face? How comforting might the event be? Do many seek something Creation has been too wise to provide? Might we find more nourishment by communing with Creation on the scale we inhabit? Is not that communion our naturally inviting calling?

(49)

The realization that life is a wondrous gift, evolved with freedom and unstained with illogical guilt, posits Creation as benevolent and enables us to think far more of what we consider God.

(50)

The first way to sanctify life is to appreciate it. It is also the first way to express our worship for whatever its Ultimate Source may be.

(51)

What is incomplete about a religion based on life itself? Certainly, there is enough practical benefit in it and, credibly defined, Divinity's inarguable presence.

(52)

The most direct, beneficial and credible worship of the Origin of life would be life-caring activity.

(53)

It is through what we can know that we can best intuit the intelligence, intent, and character of Creation.

(54)

There is not only more logic but greater merit in having reverence for life before what "the unexamined life" considers "death" than in having reverence for life after death, which may well be a contradiction in terms.

(55)

Even a single living cell is more sacred than our entire death-oriented tradition.

(56)

When we do the best we can with our lives, we do as much as we can logically be expected to do, and we achieve two goals: We live the most satisfying lives for ourselves, others, and for all other life, and we merit whatever reward the Ultimate Source of life – trusting it is logical commensurate with its achievements – may have prepared for us when this life concludes.

(57)

Does the Ultimate Source of the universe and the life that has evolved within it delight in our prostrate praise? Doesn't it seem more fitting such a Source, present or absent, would more likely be pleased, if we can please such a creative power, by our taking good care of the gift of life and doing the greatest things we have the capacity to achieve. Might such a Creator actually admire our capacity to feel pride in how we have been evolved?

(58)

A fragile and faithful few may do much but in attempting to do too much, less.

(59)

When we live in a way that the appreciation of life's greatness invites us to, we do all we can logically be expected to and merit whatever life may come after it. To agree only requires us to believe that Creation is at least as intelligent and fair as we are. Since we are its progeny, we can, I think, assume it is.

(60)

Let us devote ourselves to our life-enabled potential and thereby be worthy of life.

MATTER AND ENERGY

We might see matter and energy as the body and life of the universe, animate and inanimate. To paraphrase the scientific theory, "Body and life can neither be created nor destroyed." As a correlate, the amount of body and life in the universe is, with the exception of nuclear transformations of one into the other, constant. All that is transient are their unitary existents.

(2)

We suggest that matter and energy are the substance and life of the universe and are, in their various combinations, life animate and inanimate. This understanding lets us appreciate that everything in the universe is alive and explains the development of animate life as a perfectly natural expression of inanimate life. There is no gap. The variable is the complexity. Matter and energy express the will to live manifest in the universe in combinations that are appropriate to the environment and evolve them toward the most accomplished forms of life possible.

(3)

Energy is the universal attribute of matter and animate life is its highest form.

(4)

Matter and energy can be seen as the basic duality of or love relationship of the universe; the creative partners, mother matter and father energy, or vice versa.

EVOLUTION

I believe animate life evolves as part of the inanimate life of the universe, which I consider our natural God, and that, as the life within it evolves, the universe becomes a more complex and Godly manifestation of life.

(2)

Speculating, one wonders if we're part of the first evolution of matter and energy. The perfect plasticity of the mathematics in the ability of matter and energy to accommodate every occasion in a perfectly natural way and a myriad of other wonderfully smooth functioning aspects of life and the universe may indicate that we are part of one in a series of many evolutions. Considering the innumerable pitfalls inherent in such an enormous unfolding, we might, lightly speaking, be glad we weren't part of the first evolution. It might have been a bumpier ride.

(3)

The human race is really a delightful but often sadly disappointing excrescence of the biomass.

(4)

Creation seems to be made up of each animate and inanimate entity, evolving in its place and functioning, during its lifespan, with a complimentary duality of being and becoming.

(5)

We are the cleverest of monkeys. Were we only happy with the recognition. Then we could be at peace as a highly gifted member of the community of life on earth, with whom we share a nourishing biosphere, distinguished potential, and a common destiny.

THE ORIGIN OF LIFE

I am a scientist yet I believe everything is alive, but in two manifestations: animate and inanimate life. What is all the motion and heat in matter but the basic and advanced processes of life? This belief relates importantly to the question, "How did life originate?" I believe the question is, "How did the living universe originate – beyond its birth to the ultimate why of its origin?" Once the universe exists, we have life, at least, in its basic forms and processes. So the real question is, not how life began, but how animate life evolved from inanimate life. Since we have the substance and capabilities of matter, we know the answer is that, under the right conditions, inanimate life combines into more complex forms that we call animate life. Succinctly, animate design and life evolve from inanimate design and life.

This realization provides no conflict with the idea of a Deity. By such local evolution, the activity of our Natural Supreme Being, the Universe, is the immediate creator of animate life, while the Universe itself is either self-created or the creation of a Creator or Creators beyond our experience of it. We can be certain about the existence of Our Natural Supreme Being and arrive at detailed knowledge of it, but we can only intuit the possible existence and character of the latter by what we can learn about the former.

AN INTIMATION FROM DESIGN

The design of living creatures is so wondrous it intimates a benevolent Creation, but we can see enough of ill to puzzle the will. We finally understand the blessings do outweigh the liabilities. If they didn't, life would have ceased.

WHY IS THERE LIFE?

Why is there life? Backward, why is there a universe? Let us speculate. Since we consider existence superior to nonexistence, we have our answers.

But let's intuit more. In choosing between the existence of the universe and nonexistence, the choice, if we are willing to indulge Creation with the ability to choose, was obviously existence.

Now that we have the universe, what does it enable? All we can know. Among the multitudinous entities it enables is the opportunity for animate life to exist. So the evolution of animate life is part of the potential of the universe; therefore, the existence of it is part of the fulfillment of the universe.

Since we consider the universe to be alive inanimately, in every particle of matter and quantum of energy, we can also see how there can be what we call animate life. Under the right conditions, it evolves, through complexity, from inanimate life.

Finally, we can see that the universe is completely alive: inanimately, as matter and energy, and animately, as creatures.

(2)

There is a specious answer to the question, "Why does life exist?" The question can only exist because life does, or there would be no one to ask the question. But this response merely provides the reason that the question exists, not an answer to it.

(3)

Logic, and witty commentary, tells us that life exists, first, to be lived.

THE GIFT OF LIFE

What other gift is as great as life itself, the gift of gifts, which enables all we can know and value?

(2)

Sometimes when people say, "God bless you," I say "Thank you," but think I wish they had said, "Life bless you" and tell myself I've already been blessed.

THE GREATNESS OF LIFE AS GIFTED

Since Creation is greater than our understanding, we can infer that it has gifted life in a way that transcends our understanding, and, since we can know nothing greater than Creation, we can infer that, if life could have been gifted in a greater way, Creation would have very likely chosen it.

THE GOAL OF HUMAN LIFE

The goal that is equal to human life, as a highly intelligent and sensitive form of life, is the mutually considerate fulfillment of our finest possibilities. Devotion to it provides us with a goal that is in agreement with Creation's own devotion to the fulfillment of the possibilities of matter and energy.

(2)

We are, if logic obtains, and we believe it does, born to live – first by caring for life; then by fulfilling its life-directed possibilities. Both activities enable us to enjoy the natural mental, physical and spiritual rewards that spring from them. They also help us save ourselves from our own self-inflicted mass extinction and from the everyday death that prevails when life's potential goes unrealized.

LIFE FLOURISHES

When something is nourished more than it is harmed, it can flourish. Life flourishes. So the good that nourishes it is more prevalent than the evil that harms it.

In fact, when we look at the burgeoning human population, we could say that Creation has overdone the benevolence of our situation and our ability to improve it. But we have learned that when we suspect Creation of erring, to suspect the error is our own.

Reviewing the situation, we realize we can more accurately say that Creation has gifted human life with enough good so that we can flourish in extraordinary numbers but has also given us the intelligence to realize that we should self-limit our numbers to keep ourselves and our environment healthy. If we fail to do so, then it seems inevitable, in a naturally functioning universe, that we will limit our numbers through a self-inflicted calamity. We have the capacity and the freedom to determine how wisely we will flourish or how foolishly we will perish.

(2)

When something is harmed more than it is nourished, it can wither and pass away. But life has increased, like a plant in a fertile garden. One of our callings is to keep the garden fertile.

(3)

Recently, we see signs that the overall life of the planet and the conditions that support it may be withering more than increasing. To the degree the change is due to our early development and current carelessness, rather than to the life-support system the universe provides the earth, we can reverse it.

LIFE WITH FREEDOM

Since we detect no unnatural interference in our lives, we conclude that life has been gifted with freedom. We can make, within our capacities, the world what we wish it to be. If we elect the ethical principles inherent in the appreciation of life itself, we are most likely to preserve it and be blessed with abundant life.

(2)

Freedom is a foundational aspect of Faith In Life, because it is an expression of life as it has been gifted, in individuals, who, we trust, have been endowed with the capacity to make their own responsible decisions.

In fact, naturally functioning human life can only exist when individual freedom of choice does.

(3)

Freedom is often confounded with irresponsible behavior. Yet the continuance of freedom requires responsible behavior. In fact, the more responsible behavior there is, the more freedom distinguishes and justifies its existence, and the more irresponsible behavior there is, the more such behavior is likely to precipitate the curtailment of freedom.

(4)

Many well-meaning people who consider themselves proudly American militate against beliefs that are not in agreement with their own. But, since America was founded by people in quest of freedom, the advocacy of anything but freedom is un-American.

(5)

We respect the individual and his or her right to lead life as he or she chooses, so long as his or her intent is not to harm life.

(6)

Belief in life requires us to value individual beliefs. We may suggest insights as cordially as we can manage but we should never attempt to foist a belief on another person. I often hint with a smile and leave the perception of what I said to the other person's consideration.

(7)

Many have never considered individual freedom part of life, sanctified. Since freedom of choice is not in agreement with their own beliefs, they hope to curtail it. They are likely unaware of the harm death-oriented beliefs have done to their own lives and to all other life.

(8)

If a person we make a smiling suggestion to makes that rarest of adjustments, that is, rethinks his or her position, we may have helped save his or her life for life, which would also help save his or her life for Creation.

(9)

It is commendable to go beyond the toleration of what other people believe to loving them for what they believe, providing, of course, that it is life enhancing.

(10)

Life has been gifted with sufficient freedom for us to develop a value system that provides for our survival and encourages our fulfillment.

(11)

While many may not be equal to freedom, the abuse of freedom is preferable to the absence of it.

(12)

Why does the abridgement of freedom lead to a world that seems surreal? Natural life can only exist in freedom.

(13)

Force is more potent, at least in the short term, than philosophy and tends to exude more dignity, primarily unmerited. Note the forced suicide of Socrates, the flight of Aristotle, the beheading of Cicero, the ostracism of Spinoza, the self-exile of Voltaire, and the imprisonment of Russell. But time can rectify abuse. What is the influence of the people who exerted force against these philosophers and what dignity do they retain?

(14)

With freedom of religion, different religions may flourish in peace. Without it, there can only be irreligious persecution, which not only harms the ones who are persecuted but invalidates the religion of the persecutors.

(15)

Life is a fact and freedom a necessity of its fulfillment.

(16)

Societies that enable individual freedom do better because the free mind, which can express the potential of life, is individual.

(17)

Human potential contains, by definition, a multitude of inspirations and achievements we cannot foreknow. One reason freedom is essential is that only within it can human potential reveal itself. We can only enable the ability of individuals to have inspirations and create achievements and thereby ever more bless our lives. Imagine a human even a hundred years ago attempting to foresee what human inspiration has since wrought.

(18)

Indiscriminate disasters prompt us to understand that we are on our own and ought, instead of blaming or losing faith in a rational concept of God, accept the freedom life has been gifted with and do our best to prevent disasters and to help one another survive them.

(19)

I believe especially in the freedom of the life-oriented, considerate individual.

THE APPRECIATION OF LIFE

It is only through the rightful appreciation of life that we can sanctify it and save it.

(2)

We appreciate Creation primarily because of its life-creating and life-sustaining capabilities, which exceed, at least on earth, the amount it life-terminating capabilities; otherwise life on earth would not flourish. In fact, we consider the life-nourishing workings of Creation magnificent. When we live in agreement with Creation's life-bent intent, we intuit that our lives can be magnificent.

(3)

Life is the basic treasure of the human race and of all other creatures and plants.

(4)

In the death-culture, the appreciation of life and the rest of Creation is still primarily an individual solace.

(5)

I would like one day to visit a park, devoted to the appreciation of life, from subatomic particles to elephants. A visit to a school similarly devoted would also provide a life-enhancing experience.

(6)

The most likely way to enable world peace is to establish a worldwide appreciation of the central importance and greatness of life.

(7)

There is more brilliance in even a single cell than in all of our science, art, and, yes, our religions. Yet we still perceive no compelling reason to value life itself above all else.

(8)

My intention is to pick up this neglected orphan, life, wash it off and show it to you, so you can see its beauty, love it, and help make an inspiring home for it.

(9)

There is little that cannot be demeaned and often with superficial wit. What is far more intelligent is to appreciate the given and elevate it to a level commensurate with its inherent value.

(10)

We have always undervalued life. Our inability to appreciate its true value is the basic reason we have depredated it and now threaten to annihilate it. Life has never, in fact, been described in a way that is equal to its greatness. Even our most sacred beliefs still demean it, assuring us, "Don't be too concerned; there's something much better waiting for you after your brief stay in this vale of tears."

(11)

My own writing is my best effort to do life justice. The effort is, as the fulfillment of all of our life-oriented capacities should be, a foundational part of my religion.

(12)

Those who see value in life will more likely find value in their lives, while those who see nothing in it will more likely find nothing.

(13)

The best way to worship life and its Creator is through the way we live it. Works do wonders. They constitute a practical worship.

(14)

The human race has never had a religion equal to life, because it has never had a religion based on the intelligent appreciation of life.

(15)

We have never had a philosophy or a religion that is equal to life because we have never viewed life in a sufficiently insightful way – as the enabler of all our blessings, the ultimate good we can know, and the worthiest recipient of our intelligent devotion.

(16)

When we talk about the preservation, enhancement, and enjoyment of life as our most precious treasure, we are obviously talking about a new world and one we desperately need to enable.

(17)

There is a delightful cause and effect, or pretty predictable sequence, that generally obtains: when we are involved in an activity that is worthy of life, we feel that life is worthwhile.

(18)

The earth and the sun do give birth to and sustain astonishingly beautiful forms of life. To appreciate the spectacular blessedness we are all part of, we need only gaze at the silvery moon. In fact, the more we see of the rest of the solar system, the more we appreciate how miraculous life on earth is.

(19)

In a life-appreciating world, there can be no credible concept of nihilism; it is an aspect of the inevitable emptiness wrought by the realities of the death culture.

(20)

To stand on the stable earth and, through the transparent air, see the great gifts of Creation and the lovely works of man, woman, and other creatures – the entire panoply of life in what we may consider, at such blessed times, the playground of the earth – and to know it is your time to be part of all the vibrant wonder, a sensate participant of its intelligence, fulfillment, and potential for very likely 80 years or so – this is joy, exultant, irreplaceable joy!

(21)

I think it is a more logical and certain reverence to the directly unknowable Origin of life to substitute the word "life" for "God" in many instances; for example, if one day we said, in appreciation for something good that just happened, "Thank life!" Such an event would be a small reassurance that some of us are getting our priorities right. But will we ever hear, at least, as an acceptable variation, the official use of the phrase, "In life we trust"? The immediate answer reveals how far we have to go to turn our eyes from the empty sky to the miraculous blessings before us.

(22)

Look insightfully at the present doings of humanity worldwide and you will detect, more often than not, that the precipitant of most human misery is irreverence for life.

(23)

Let me imagine awhile the basic symbol of life. It might be inspired by matter and energy, which are, of course, the basic components of the universe and so of its animate life. Subatomic particles seem a bit nonspecific. On the other hand, the atom presents itself as possible, since, under the right conditions, it combines with other atoms to create molecules and then more complex forms of life. Atoms can be seen, in fact, as the fundamental natural life creators of the universe. How about a sperm and egg? Or a human cell? How the appreciation of life opens up a whole new world of wonder, yet to be described and illustrated!

(24)

Life is the greatest creation we can know. Let us believe in it more than the life-rejecting ideas we have invented.

(25)

One of the most challenging aspects of writing about life itself as a modern religion is never erring in a way that makes my reasoning suspect. In fact, I try to make as much as I can of what I say self-evidently credible. So dare I ask if life *is* our heaven – demeaned, denied and depredated even as we inhabit it?

(26)

We are still, most of us, in the midst of life, strangers to it. Yet many of the consolations and joys of life grow out of simply appreciating that we have it. Without life as our natural foundation, purposeless and bereft we may well feel.

(27)

To teach people to feel guilty about their perfectly natural desire to live in harmony with life's promptings, which of necessity includes mutual consideration, bespeaks an ignorance that begets irreverence for life.

(28)

Recognition of the self-sufficient greatness of Creation's design is part of the natural appreciation of life.

(29)

We should appreciate life with all of our capacities – mental, physical, and spiritual. How exhilarating and saving the experience!

(30)

Let us now remake ourselves toward life. Let us stop rejecting it because we perceive imperfections and it does not last forever. Let us, instead, embrace life as it has been gifted and let our unanalyzed views of its imperfections, which are ours to improve, and our overriding hopes for immortality recede into our life-insensitive beginnings. We need not do so as a Camus "rebel," who vows to live in spite of death, a self-imposed, strident struggle, but as people who are at peace with life as it has evolved, because we truly understand its inestimable greatness and know we are privileged to be part of it.

LIFE AS A NATURAL MIRACLE

The real revelation is that we have discovered, through our scientific understanding of it, the greatness of life.

(2)

The true miracle is life itself. It reduces all the others we have reports of to childish tricks.

(3)

See the miracles of life before you, even the wonderful agility and intelligence of the smallest creatures.

THE SANCTITY OF LIFE

We have always looked for a unifying principle for all of our disciplines. Religion, philosophy, science, art, government, and business can discover it in the legal principle of the sanctity of life, enlarged to its rightful scope.

(2)

The sanctity of life must be based on the sanctity of the individual, who is the very embodiment of life, and must include, for natural ethics to exist, the freedom of the individual to make choices, hopefully wisely, which would include consideration for one's own life and for all other life.

(3)

All life is sacred, along with all that originates, supports, enhances, and helps fulfill it.

(4)

The sanctity of life includes individual, mutual, and general sanctity; it is a complete ethic. How can we enable its potential? By understanding and committing to the miraculous and irreplaceable existent life truly is.

THE ACCEPTANCE OF LIFE

Let me dare, for once, to say it straight out: by accepting life, we can make it a new heaven; by rejecting it, a continuing hell.

THE CARE OF LIFE

Life, we find, is in our trust and needs our most devoted care.

(2)

The better we take care of life, the more reasons we will have to live for it and be satisfied to have lived.

THE WILL TO LIVE

Each person's will to live is an expression of the will to live of all Creation. It constitutes each person's share in the universal will to live, which can be considered naturally divine or naturally so with, perhaps, a supernaturally divine component. It is an expression of Creation's devotion to life, and can be an expression of our own agreement with Creation.

(2)

Nietzsche's "Will to Power" is a heavy-handed and self-aggrandizing concept. It is not in keeping with a grace consonant with the delicacy of life or the necessities of evolutionary survival. We have, more exactly, a Will to Live, which has individual and mutual aspects and is in agreement with, in fact, part of, the overall will toward life that we see in Creation's determination to establish life in whatever nook and wider locale it can.

LIFE-ENHANCING CONTENT

We define our lives, to a great extent, by the content we choose for them. Life merits the most life-enhancing choices we can make.

LIFE REWARDS

The quest for material excess is partially based on the general unavailability of natural life rewards – mental, physical, and spiritual – which a life-oriented society would provide in abundance.

(2)

We are able even now to moderate the excessive quest for material things when we realize that, with comfortable life support, we can be free to concentrate on the activities that provide us with the natural life rewards we seek, such as the joy of achievement in areas we feel we have talents for, love relationships and families supported without anxiety, and fellowship uninhibited by economic concerns.

(3)

When we measure our lives in intelligent joy, as the overall effect of fulfilling our own capacities and helping other people and other kinds of life fulfill theirs, we are much more likely to find life delightfully and sufficiently rewarding.

THE BIOSPHERE

The entire biosphere is a sacred place – sacred to life, the natural universe, and Creation's Ultimate Source. Since we assume a benevolent Creation, and thereby a worship-worthy one, we also assume the biosphere is intended as a place of intelligent fulfillment, joy, and, if we wish and in our own ways, of worship.

We should utilize the resources of the earth freely but we should also realize that it is part of the body of our natural God and a living representative of a God or Gods beyond our experience, and that it works tirelessly to cleanse itself and remain beautifully vibrant. We should work as diligently to preserve it, knowing even dust is sacred.

(2)

The scars of our past misbehavior are on the face of today's biosphere.

(3)

The life of the earth can be seen as a great, vibrant being that, with sufficient care, will flourish throughout its lifespan but, with enough mistreatment, may sicken, fade, and prematurely die.

(4)

If the environment is not as clean as it once was, can we be as healthy as we can be, and, if the air is not as pure, can we, whose minds are nourished by oxygen-rich blood, be as alert as we can be?

(5)

If the life of the earth were a leaf, it would have brown edges; a tree, dead and dying branches.

(6)

In evaluating global warming, pollution, or over-harvesting, the overall effect is not that a particular environment or species is threatened or is already extinct. It is that the overall life of the earth is dying under the cumulative effect of billions of human beings, most of whom are indifferent to the sacred value of life and consequently insufficiently inspired to adjust to living in an eco-loving manner.

(7)

What has created human life on earth? The environment. It is, at least, the immediate creator of it and, as part of the natural universe, the proximate Creator that we know exists. As such, it merits our care and reverence.

(8)

Sun-resplendent and rain-graced days are two wondrous, alternate life-nourishing gifts.

(9)

The health of the earth is now very much at our mercy. We know we live in freedom. So Creation evidently trusts the capacities we have evolved with and has the patience to let us realize them. Failing, we can relegate ourselves to being a destructive species and Creation may evolve another one more capable or better adjusted. While its evolutionary intelligence obviously dotes on each creature it evolves, it is and rightly ought to be in love with life in general more than with any species.

(10)

All has been given to us – life and land, water and air, space and time, intelligence and prowess. Yet we are not satisfied.

(11)

Keeping life healthy and beautiful is not only a vital way to preserve and enhance our own lives; it is also a vital way to help keep the life of our Natural Supreme Being, or what we call the Universe, healthy and beautiful.

(12)

When one realizes how faithfully the environment of the earth has nursed up life from matter and energy and how long it has sustained it, one realizes that, even given storms, quakes, and partial extinctions, it is, overall, incredibly benevolent. In fact, I know of no activity, human or otherwise, that can compare with its steadfast care.

(13)

Yes, the biosphere has destructive severities, but it has been supportive enough to evolve life and sustain life for millions of years. How many things are we aware of that are so faithfully nourishing? Will be ourselves be, in our time, as faithful to life?

(14)

We stand on and live off the living, giving earth.

(15)

Reverence for the land, water, and air is a fundamental component of Reverence for Life.

(16)

It seems entirely appropriate that we revere and, if we wish, worship the natural wonders of Creation. We have no more direct communion with their Origin.

(17)

Imagine how delicate and potentially evanescent the atmosphere is. Yet it has adhered to the earth and supported life for billions of years.

(18)

While the atmosphere is miles high, the range that animate life can inhabit is very near the surface of the earth. In fact, most of the animate history of the earth has taken place within a few feet of the ground and most of human history within perhaps ten feet of it.

(19)

As the human race grows to untenable numbers – at least, until we cultivate the earth with far greater care – we notice the decreasing plenty of the other forms of animate life. Our inability to adjust to our place in the life of the earth indicates, I think, that we do not yet have a concept of it that is sufficiently compelling. Will only mounting debacles awaken us? Are we a species that will not awaken? Will we, in fact, extinguish ourselves?

And what, my friends, is more likely to save us, reverence for death and the solemn expectation of our universal end, or reverence for life and the enlightened expectation that we may enjoy our natural human lifespan?

We watch the waning and wonder what each day will reveal.

OUR NATURAL PARADISE

The earth is, astronomically speaking, part of the heavens. We have only recently been able to appreciate observation of it from elsewhere. So welcome. You are in Paradise, at least, the most paradisiacal place for millions of miles around.

(2)

The view of the earth from space helps us understand that Creation's gifts have been abundant and that the continuing care of the earth and the life on it are now primarily up to its inhabitants – a condition that seems both logical and just.

(3)

Our ancestors, even in their most ecstatic visions of heaven, could not imagine the material and medical well-being we have been able to attain through the capacities inherent in human life as it has evolved.

(4)

The earth can be appreciated, in perfectly natural terms, as a heaven of life – an effulgent grace in the divine body of the living Universe.

(5)

We might see the earth, in visual synecdoche, as a natural Eden with a computer in it.

(6)

We can, with enough wisdom and effort, rejuvenate the earth and experience again its natural plentitude of clean air, pure water, verdant land, abundant wildlife, and the naturally inviting Eden they comprise. "Heaven on earth" can now be considered an achievable goal.

THE BLESSINGS OF THE NATURAL

The perfectly natural functioning of the universe we can know is an astonishing blessing; it enables norms on which we can rely.

(2)

When we accept that the world we experience functions in an entirely natural way, we are free of the phantasms that have haunted people for eons. They are biological impossibilities that have functioned as diminutions of life as the glorious natural and unencumbered gift it is.

(3)

The more we appreciate the wonders of the natural world – which is, in a word, miraculous – the less we require the consolations of a so-called supernatural world.

THE STABILITY OF THE EARTH

Is the earth a stable home for life? Yes, enough to have evolved and sustained the life on it for eons. So amid the storms, quakes, volcanoes, and asteroids, we have every reason to maintain our faith in the earth and its vibrant life.

NATURAL GODS AND GODDESSES

When we cultivate our intelligent and compassionate capacities to preserve, enhance, and be gratefully happy in our little Eden, we may see ourselves, very appropriately, as the natural Gods and Goddesses of the earth. The greatest challenge to the distinguished vision is for us to understand that what we regard as death is the natural completion of the gift of life and to be satisfied with our share of life, to understand the wisdom – in fact, the grace – of lifespans and be content with our own. Then we may reevaluate life, rescue it from the grip of death, and realize our most distinguished and sacred potential. We may say that heaven, after all, does await us beyond death, but the way to it is enlightenment while we are alive.

THE OTHER ANIMALS

All creatures merit their own sacred place in the panoply of life. To confirm their worthiness, we need only appreciate with wonder their true capacities.

(2)

Simpler creatures offer a wonderful example of the sane conduct of the basic activities of life, and we, who cannot only think, but over think, can learn from their unexceptional dutifulness.

THE NATURAL UNITY OF LIFE

The unity of life is more than a spiritual intuition. All the inanimate life in the universe originated from the Big Bang and the animate life from the synthesis of it on each planet. So all life is truly one, while each of us is an individually precious embodiment of it.

(2)

Our invented divisions separate us unnecessarily. As we realize that our well-being is very likely in our own hands, at least, in this life, we will be able to feel the natural sympathy for one another and, in fact, with all other creatures that life invites us to share. Then we will be more determined to enable the natural unity of all the animate life of the earth.

(3)

The substance of the universe is one soil and its life, animate and inanimate, one life.

FULFILLMENT

The fulfillment of life is a basic expression of a commitment to life and a determination to realize its greatest potential. It is a foundational part of a religion based on life itself.

(2)

The encouragement of naturally balanced fulfillment is based on our faith in the capabilities of humans, as they have evolved, and therefore in the capabilities of Creation, as it has evolved us.

(3)

I think we can infer from the way life seems to have been gifted – miraculously capable with patient silence – that all Creation asks of its creatures is appreciation of life's gifts and considerate fulfillment of its potential.

(4)

All Creation tends toward fulfillment, individually and as a whole. We should live in harmony with the tendency.

(5)

Since Creation has provided us with the gift of life, it is our greatest benefactor. Though life-directed behavior, we can be our second-greatest benefactor.

(6)

The self-sufficient conduct of life has two necessary components: comfortable survival and considerate fulfillment.

(7)

The fulfillment of one's life requires consideration for others, not only because it is ethically right to consider the needs of others, but also because our needs are not only individual but mutual. If we attempt to enjoy our lives without mutual consideration, we deny ourselves many fundamental fulfillments, because we reduce our genuine participation in the lives of others and their participation in our lives. We also diminish our self-esteem and the esteem in which others hold us. So, overall, we reduce our own enjoyment.

(8)

While insensitive individuals may prefer only self-interest and even pride themselves on it, the very insensitivity that they require for the negation of their mutual needs reduces their possibilities for fulfillment, whether or not they are aware of the loss.

(9)

While a certain amount of self-doubt, especially in quest of a daunting goal, is understandable, a disabling degree of it constitutes irreverence of our own lives and for the possibilities of Creation only we may enable.

(10)

We must always be on the alert for subtle suggestions from the death culture that attempt to block our efforts to live satisfying lives.

(11)

We should attempt to fulfill our finest potential, not as an anxiety-encumbered burden, but as a relaxed ideal. Then we may live completely and enjoyably.

(12)

When a life is wisely goal oriented, it is like a tent with a pole. It has a definite shape and strength. Without a central goal, it can collapse into an indeterminate mess.

(13)

The fabric of each life is part of the fabric of all life. A fulfillment graces each life and all of life, while a loss diminishes each life and all life.

(14)

When we appreciate life, we grow to love it. We become in touch with its gifts and understand its needs.

(15)

The fulfillment of life-oriented human potential leads to the realization of our greatest good and happiness.

(16)

When we see a dynamic individual or society make outstanding achievements, we should be generous in our praise. Compliments are, as we all know, rarer than criticism – a subtle expression, I think, of our diminution of life, instead of our elevation of it.

(17)

When we allow life to exist in intelligently rewarding ways, we balance, with the natural way life has been gifted, the pangs of its inevitable conclusion.

(18)

There is no higher destiny we can know than a life greatly and considerately fulfilled.

(19)

When people seek to trivialize their lives with an excess of superficial enjoyment, we might ask, "Fun? All you want to have is fun? Noble people aim at greatness!"

But we ought to be disinclined to be severe and encourage life-oriented fun in plentitude. After all, joy graces life and encourages its correct valuation, while somber cautions traditionally inhibit our natural right to abundant delight.

I also believe that in the pursuit of greatness there can be, along with the possibility of extraordinary sacrifice, exquisite fun.

(20)

The truly noble among us can imagine no greater distinction than aiming at greatness – the fulfillment of our finest possibilities, including our most helpful services, regardless of the rewards or punishments we may encounter.

(21)

We can each ask ourselves, what is the quality of my character, noble or ignoble – or, more generously, the preponderant aspect of it?

(22)

Fulfillment is the way we make our capacities part of the living experience of the Universe and the expression of our gratitude to it and any Creator or Creators beyond it. We gift back to the universe, or to any Creator or Creators beyond it, a fulfilled life.

(23)

If our natural God, the Universe, could speak, other than through its creatures, I think the way life has been gifted lets us infer that it would say something along these lines: "I have evolved and sustain life. I devote my substance and energy to it. Devote yourselves to it and you will have the same purpose I do. Understand the gifts of life. Love, care for, and enjoy them – for yourself, other people, and all other life. And remember: your life is part of my life and the quality of it. So do the best you can with your share of it.

"Also, when your time to experience life attains its span, I hope you will have lived enough to be grateful that you have been able to enjoy the gifts of life and can let your life reach its completion with inner peace and satisfaction.

"I would also like you to know that, since your life is part of my life and you have lived, you will always be part of how I have lived."

If there is a Creator or are Creators beyond the natural universe, I think they would say much the same but might modify their words as follows: "We create and sustain life, because we value it. Our approach is to enable the conditions for life to evolve and flourish. We hope that you will recognize our purpose in what we have created and make your own purpose consonant with ours. And remember: since your life is part of the life we have evolved, your life is part the quality of it. We hope you will fulfill the finest possibilities of your life out of reverence for it and to help fulfill and distinguish all we have created."

What greater calling do we require?

(24)

The wise management of our own lives, particularly in times that test us, encourages us to keep the wholeness of our lives ever before us, so that we do not subordinate it to any part of our lives – whether a love, a career, a talent, or any other aspect, particularly when a difficulty grows out of any part of it. To do so is an injustice to our whole lives and may be seen as creating a handle by which the world can shake our entire beings.

(25)

What about the injunction to live for the moment? It is the perfect prescription for an unsettled life. Each moment, while vibrant with possibility, is only a passing part of our lives and never consequential enough to negate the wholeness.

(26)

Prudent moderation of our needs can enrich our lives, especially at times when there might otherwise be little but empty days, with intimations of desuetude moving across our souls like desiccating winds.

(27)

Endless need can never know satisfaction. Among other deficiencies, it negates the value of other individuals, so we may never allow ourselves a completely satisfying relationship.

(28)

Turn the pages of your life until you come upon a page that has something special written on it; you need not look for two.

(29)

If we deny the thoughtful fulfillment of our own lives, who may we blame for our loss?

(30)

We are at the junction of a new and necessary transformation from death-orientation to life-orientation. Death-orientation is still pervasive. So it is especially valorous to accept life and vow to live the greatest lives we can.

(31)

When great numbers of people are convinced that life is a blessing we may enjoy, rather than a curse we're condemned to endure, we will, of course, find it easier to improve the conditions of life and actuate its fulfillment and joy.

(32)

Let us be wary of remonstrations about perceived shortfalls between life as it is and as it might be and commit to its excellence. Let us also be aware that shortcomings that seem inherent may be based on our lack of understanding, while ones that are caused by human misbehavior invite our efforts to help remedy them.

(33)

If you enjoy your portion of the earth's animate lifespan, you will enjoy the fundamental gift that all people, for all time, are invited to enjoy.

(34)

It is perhaps reasonable to expect human possibilities and fulfillments to expand along with the universe or, more simply, in time.

(35)

There is great joy in mutually enabled desublimation, but a mutually considerate society requires a certain amount of sublimation.

(36)

Each person's birthright is a life that can be fulfilled in its natural proportions, which includes one's own potential for life, as well as the potential of others for life.

(37)

Did I live or die today, I ask myself, and, if a bit of both, in what proportion?

(38)

In the Republic, Plato tells us, "We may call that part of the soul whereby it reflects, rational; and the other, with which it feels hunger and thirst and is distracted by sexual passion and all the other desires, we will call irrational appetite, associated with pleasure in the replenishment of certain wants."

While we remain astonished at his accomplishments, we may differ and maintain that it is rational to wish to eat, drink and make love, usually moderately but occasionally immoderately; in fact, that it is irrational not to want to eat and make love, since we need to, are born to, and our very survival is dependent on the activities. So our rational fulfillments actually depend on our physical fulfillments.

(39)

What delight there is in sentience – self-awareness and reason, creativity and memory! But also what delight there is in sensation – our overall feeling of well-being and the joys of the five senses, lovemaking and dining!

(40)

Excess in pleasure has a limit, because it leads to exhaustion and ennui. It appears that we have a natural potential for pleasure in each area of fulfillment and that we find the truest satisfaction between abstinence and overindulgence. We return to Aristotle's *Nichamachean Ethics* and moderation in all things – but with, of course, an occasional fling.

(41)

Self-realization, considerately achieved, is life's greatest distinction and joy.

(42)

Set great goals for yourself and fulfill them with confidence and joy.

WELL-BALANCED LIVES

Are we meant to lead well-balanced lives? Let us look at our exteriors. Are they well balanced? Does that not imply that our interiors ought to be, too?

(2)

The majority of nature's creatures behave in well-balanced ways, and we assume their behavior implies that nature's most intelligent creatures should behave that way.

POSSIBILITY

The waves of possibility, cascading against the shore of existence, have placed a new world at our feet, if we will only see it, pick it up and treasure it.

ACHIEVEMENT

It is a distinguished achievement, perhaps the most distinguished individual achievement, to live with placidity, thoughtfulness, intelligent effort, and natural enjoyment, generally untroubled that our lives will in time complete and confident that by so living we are doing all that we can to be worthy of life and all that can rightly be expected of us.

(2)

Since we can fashion inventions out of matter and energy, we are Creation's animate enablers of many of the possibilities of matter and energy.

(3)

Great achievements flow most expectedly from great minds, bodies, spirits, organizations, and societies.

(4)

Creation enables challenges that plum our intelligence, physical prowess and spirits, and it has provided us with the minds, strength, and fortitude to deal with them. In fact, the very challenges are what enable our aptitudes and distinguish our efforts. We have evolved with the equipage to triumph over our challenges; otherwise, we would not have survived.

There is unfathomed profundity in the dynamic – life enabled, challenged, distinguished, and complete. There are inherent in it trials but also abundant intellectual, physical, and spiritual rewards.

We are, in fact, privileged to partake in a challenge unsurpassed – in our very awareness of it and our efforts, successful or not – in the wide universe. Participation is the privilege of animate life, wherever it chances to evolve.

(5)

To help guide our individual efforts, we have the capacity to assess our own achievements anytime we wish and until our last moment. Yet we must finally leave the assessment to others. For instance, how accurately could Mozart, dying in debt, or Van Gogh, unsold, assess his contributions?

(6)

There are deep natural rewards in laurels merited, inner diminution in those unmerited, and just rebuke of others for those merited but denied.

(7)

Each praiseworthy human achievement helps enable life and invite our faith in it.

(8)

In the quest for achievement, we hope for the wisdom to lead well-balanced and happy lives.

(9)

An award is not distinguished by its trophy or other content but by those on whom it is conferred.

(10)

We strive to be our greatest selves because we know life merits our achievement and because our own lives and the lives of others will benefit by such conduct.

(11)

The human race will be what it achieves.

(12)

Let us each strive to make life great for ourselves, all other life, and for all Creation. There is no higher calling in or out of the cosmos.

SUCCESS

Creation has been successful to the extent that we have evolved in an environment that continues to sustain us and are remarkably capable. Let us be successful with the goal of doing justice to Creation's achievement.

(2)

We may have many kinds of success, some spiritual, others commercial. While spiritual success is often deeper, there is also unique value in commercial success. Among other things, it supports us so that we can enjoy spiritual rewards.

(3)

There is only one alternative to a spiritual or commercial success and that is a spiritual or commercial failure. But then what is success? That we do our best, taking care of the causes of doing so, and let the effects be what they may. Ours is to do what we can; others to judge what we have done.

LIFE AND DEATH

The human race will be happier when it is at peace with life and its natural completion and our real challenges come from the care, betterment, and enjoyment of our lives.

(2)

If we weren't over-concerned with our mortality, how would life look? The discovery will, I think, be a global delight.

NATURAL SALVATION

Fulfillment, intelligently and considerately achieved, enables the natural salvation of one's life, the lives of others and peacefulness at life's completion.

EACH LIFE SOUL

Each life, as a vibrant compound of matter and energy, is part of the life of the universe, and each life soul is part of its life soul – an individual will to live that is an aspect of its universal will to live.

(2)

The naturally vibrant spirit of each living thing is part of the naturally vibrant spirit of the Universe, which we consider our natural Supreme Being. So we consider each life part of the life of our natural God. We do not know whether each life is also an expression of the will of a God or Gods beyond the universe and assume we ought to be content with what we can know.

A NATURAL AFTERLIFE

Everything we do is forever enshrined in the temple of time. Since it is something that has happened, it can never not have happened. In a very natural way, existence includes eternity.

(2)

When you live, you do not die but save your life; and, at the end, rather than die, your life completes.

(3)

Let us deal with the hope for an afterlife in the traditional sense. If a God created life, is involved in our ongoing lives, and has prepared an afterlife – whether the God is the universe itself or a God or Gods outside of our experience – such a God would, logically, love those who take care of and do their best with this life, since it is what such a God has created as our initial experience. So, should there be a life after this one, I believe we who have Faith In Life and act in consonance with it will be welcomed into it as readily as anyone else and perhaps more readily.

(4)

What other conduct is as praiseworthy and sacred as the care and fulfillment of life? Any proposed alternative requires that we be, to some extent, dismissive of the worthiness of this life.

(5)

What if our lives are finite and we cease to exist when our bodies can no longer create our life energy or what we call our souls? Do not our lives gain in value – in fact, achieve their rightful place – by being appreciated as all that we truly have?

(6)

The belief that this life is our only certain experience does not demean it but distinguishes it. Let us imagine each of our souls, or spirits, as if it were the glow of a flashlight. The illumination is present as long as the batteries have energy and the bulb a wire that is intact. When the energy is expended or the wire breaks, the light goes out. Yet all else that was there when the light shone is still present. The flashlight appears "dead," but nothing since its extinction has departed. If such an end is also true of our souls and they do not pass on to another life, should we lose faith in this life? No, we should have more faith in it, because it is all that we have.

(7)

If an after-gift awaits us, other than the balm of absolutely nothing, I will be as surprised and happy as anyone else. I also believe my Faith In Life and the conduct it has inspired will allow me to be as welcome as anyone else, if conduct in this life influences how we are welcomed.

(8)

We can think but shall sleep; talk but be silenced; achieve but be stilled. Helpless against mortality we are, and nothing makes the point more poignantly than reading the thoughts of bright people whose lives are now complete. Yet their thoughts, words, and achievements do live on, in that they happened and, as a result, can still happen in our own living minds and enhance our lives. Their expressions and achievements are among the most distinguished aspects of life's lasting legacy.

(9)

No vibrant sinew has ever been strong enough to withstand the tears of time, but it is, I know, far better to be sensate and fragile than, say, insensate and eternal.

(10)

No matter what each of us does, we all live under mortality's net and so must finally "die." But we do also live. What sounder way to adjust to both certainties than to live as wisely, happily, and helpfully as we can. By doing so, we may live in the most life-consonant ways and thereby enjoy the most fulfilling lives and show our gratitude to life itself and to all that has made life.

(11)

Seeing ourselves as mortal invites us to feel more empathetic toward one another. We know we are each human and fragile, but heroic in achievement and to be embraced in tender understanding of our mutual mortality.

(12)

We list ways we may diminish death but know our completion still awaits, yet we do not live as a thankless slap at life, but with a grateful smile for it.

(13)

In the billions of years ahead, we can quite soundly assume that one day all of human life will have vanished from the earth and perhaps from the universe. Yet what we will have done with our gift of life, all that we will have achieved or ruined, will abide as what has occurred as our portion of the great unfolding of life in the universe.

(14)

Nature tries, I think, to tell us that living is our gift and life's completion a terminus, with such ensigns as ruddiness vs. pallor, warmth vs. coldness, mobility vs. immobility, and renewal vs. decomposition.

(15)

Lifespans are primarily evident to us in animate unities of matter and energy. So our understanding that, in the world we can know, matter and energy cannot be created or destroyed, except in a nuclear transformation of one into the other, has religious implications. The matter and energy that make us up are eternal. Let us do justice to our individual share of their wonder while we may.

(16)

The relationship between a rightly situated and sized planet and an appropriately nuclear star creates a certain climatic potential to evolve and sustain life. The resources of both the star and the planet are enormous but finite. We are each privileged to be part of the potential for life created by the nourishing relationship between the earth and the sun. Let us merit the grace by how we choose to live and then live on in the eternal natural effects of our lives.

(17)

Our only certain life occurs before what is traditionally considered death. Live it wisely and thereby enjoy and help sanctify it.

TIME

Time is the name we give to the interval the universe provides every creature, thing, and itself for fulfillment.

(2)

Time is not our enemy but the steadfast span that creates our opportunity to live. Our usual conception of time relates to how we divide the opportunity.

(3)

While the overall span of the universe is an eternity in our concept of time, every moment within it is unique – to each of us, to the people we love, to all other people, all other creatures, and, in fact, to the entire universe. Without placing an undue burden on our fulfillment, we ought to be aware of the unique value of each moment, month, and year so we may more appreciatively and capably conduct our share of them.

(4)

Time takes its time but achieves its goals.

(5)

All things are followed by other things, yet all things forever have their place.

LIFESPANS

Each life is made up of conception, dependent development, birth, independent and social fulfillment, and completion. When we make a healthy adjustment to the process, we accept Creation's invitation to life and can live in grateful harmony with it.

(2)

New lives begin and complete every moment, while each of us has a lifespan that is part of the overall lifespan of animate life on the earth and in the vibrantly life-creating universe.

(3)

We arrive as astonishing emanations of the essentially eternal gene pool of animate life on earth, derived from the first spark or form of it, grow into adults as we fulfill, to the extent we are able, our possibilities. Then our lives as uniquely individual beings complete, at least, the aspects of them we can know for certain, and the atoms and molecules that made us up return to their ongoing and essentially eternal life-partaking possibilities.

(4)

Birth and completion each take a moment; all the rest is life, invitingly individual and graced with promise.

(5)

When life seems shorter than we wish, the perception can inspire us, not to turn away from it as ultimately unsatisfactory, but to value more each moment and year the gift of life is ours.

(6)

Love each birthday; the alternative of not having one is not as inviting.

(7)

Creation has made the hard decisions necessary for life to flourish and has decided that each life should have a span and, as a corollary, that new life is more precious than the eternal continuance of an individual life, as well as of any species, planet, or even the universe itself, at least, in its current span. A choice had to be made or else there would finally have no room for new life.

Yet there may be a subtle promise in the multitude of planets that can likely support animate life. We are early humans on our home planet. Perhaps about the time we learn to travel to other planets, we may unravel the biochemistry of animate mortality.

(8)

Perhaps our frequent desire to live longer indicates that we have the ability to achieve longer, if not essentially eternal, lifespans, and perhaps one day the liabilities of achieving them will diminish because we will have become populous among other planets. But all of this speculation arises from the great perhaps and, just as previous individuals and generations have had to adjust to their place in the unfolding of human life, we ought to adjust to ours – and, in fact, be joyful that we are alive and able to do so.

(9)

The earth will likely close its human chapter long before the book of its animate history. Yet by caring for the earth and its animate life and fulfilling our best possibilities, we may write our finest pages, which will forever be enshrined in the earth's history.

(10)

An aspect of the natural probability of life as it has been gifted is that lifespans are of various lengths. The variability is a truth we must adjust to, even when the brevity of a loved one's life tries our Faith In Life. We must return, as we can, to our gratitude for the life, brief as it may have been, and our belief that the abiding value which invites our reverence is that there are lifespans at all.

(11)

The alternative to lifespans would result, at this time, in overpopulation to the extreme of human congestion and a diminution in the value of each lifespan in general and to individuals. We would be reduced to wishing more people would just reach the end of their lives, and each year of life would no longer be as precious. In eternally continuing lives, many would, in fact, be reduced to ennui. We would also feel guilty about the newborn lives our own essentially eternal lives are excluding, along with the fulfillments germane to infancy and youth. This enumeration is not intended to actuate feelings of guilt about the desire to live longer, or even to be immortal, but to accentuate the merits of life as it has been gifted.

(12)

It seems that all forms of animate life, from a cell to a planet, are part of the cycle of birth, fulfillment, completion, and the birth of new life – all laudable graces of matter and energy, variously compounded, separated, and united again.

(13)

A life of natural fulfillment and considerate joy cannot rightly be overshadowed by the moment of its completion and the eternity of its having existed.

(14)

Life is a gift we each have a share of. When we see a newborn or a child at play, who would not agree with the wisdom of the arrangement?

MORTALITY

Life is the precondition of mortality. Stated another way, mortality can only exists because life does. So the primary intent of Creation is, we take it, life, and mortality one of life's effects. Therefore, let us act in consonance with Creation and make life our primary intent, even though we know we are mortal.

(2)

Mortality casts a wide net, and we all live under it. Nothing, even the great universe itself, can escape, at least in its present expansion. No matter how brilliantly we think or heroically strive, in time mortality's net will lift us out of life. Yet it is the fish in the sea that enable the fisherman, not the fisherman the fish, except for the Ultimate fisherman, the enabler of both.

(3)

I do not gamble, except as I wake up every day and do my best within life's natural probabilities, and I think gambling metaphors are undistinguished; Pascal's gamble that we have nothing to lose by faith has always struck me as such. Yet such a metaphor occurred to me, and I think it has enough clarity to mention.

In the game of life, mortality may hold all the aces but we may have, at least, a full house and in time a royal straight flush, the most winning hand, by doing our best with life and leaving to mortality only the few chips we can't take off the table.

(4)

Matter and energy are more mortal in their animate combinations. Since the combinations are sensitive, they are, despite their vibrancy, relatively fragile. Does our sensitivity predispose us to mortality? Then, to the extent that we prefer sensitivity, we ought to find it easier to accept our mortality. Would we rather be insensate but immortal?

Now, to speculate wider and then, as we should, scurry back. What about any Creator or Creators beyond the universe? Of all existents, they would be sensitive. So they must be fragile. Then how can they be immortal? But now we're viewing them, as we must, with human references.

(5)

There is great dignity in doing our best in life while we calmly accept the inevitability of our mortality.

LIFE'S COMPLETION

A funeral can be what people decide to make it – sorrow at life's natural completion, a celebration of the life, or, I think most fittingly, a thoughtful mixture of both.

(2)

When a person's life completes its span, old or young, much of the individual's life remains within us as memories and outside of us as achievements. Moreover, the person's entire life, even the smallest thoughts and actions, will exist for all time, because his or her life happened and can never not have happened. Amid the loss, the wholeness of all life, which made possible the individual life and our relationship with him or her, also abides.

(3)

When we grieve over the conclusion of someone's life, we may, in swells of anguish, sigh, cry, and even utter words that may be addressed to the person whose life has concluded and to powers we know not of, even though we doubt the person or the powers can hear us. Such expressions are a natural part of the experience of loss. At such times, let us, to the extent we can, dwell on the life and the great gift is has been.

(4)

It is, I think, much more dignified and consoling to stand before the mystery of eternity with the sorrow we naturally feel at the completion of the life of someone we love and to feel, within the sorrow, gratitude for his or her life, while we allow Creation to be its own answer about what we know not. Certainly, it is more in keeping with solemnity than to have our feelings disturbed by answers for which we have no respectable foundation. Let us turn in loss primarily to our natural solace, which is life, and, secondarily, if we feel the need, to our infinite hopes.

(5)

When alive, we are all different; when our lives are complete, we are, in the stillness, much the same. But we each do live on very significantly in the unique legacy of our deeds.

(6)

As my life comes to its completion, I expect to think, "I have lived, I am grateful that I have had the opportunity, and am content. I leave the gift of life to those who can still experience it."

(7)

A consolation each of us has for the inevitability of life's natural completion is that it is part of the process that enables newborn life.

(8)

When we believe in life, we accept its total panoply – the joys and sorrows, satisfactions and denials, grandeur and degradation, initiation and completion.

(9)

The completion of our lives seems to mark the end of our individual existence, at least, as we, who are still alive, can know life. It seems to be, if you will, not a semicolon but a natural period. The recognition of our probable terminus enables the basic, life-oriented wisdom that the death culture often labels as vain and less than valorous: Love your life and do your best with it. Such conduct is, logically, our duty to its inexplicable greatness, as well as to its Ultimate Source.

(10)

Life's completion is the end of our natural experience, so there is really only one natural, personal experience: life.

(11)

Since the completion of life marks the end of perception, it cannot, as Epicurus noted, be experienced.

(12)

The completion of a life can only be experienced to the extent that the event is observed by others. Even if the completion of this life were an awakening into another life, it would not be experienced as a terminus but as a transition. So life is really the only experience, and what most fear most they will never experience.

(13)

Life is the main instrument we have to deal with our mortality. When we live fulfilling lives, we may live as if death, in the sense that all ends in nothingness and all is for naught, does not exist – because it does not. We have saved our lives from such an imaginary nothing-nothing by living them.

(14)

Graced sufficiently with life, we may live without the fear of its completion.

(15)

When the life of someone we love concludes, we can feel outside of life – as if we ourselves have been removed from it by our attachment to the person. We need time to adjust to our loss, and then we can reattach ourselves to life. The transition becomes more navigable when we reflect the grace of having had the person in our lives.

(16)

When a person I know reaches the end of his or her life, I begin to assemble my memories of him or her and, by so doing, build within myself a way for him or her to live on in me and a living memorial.

(17)

When someone's life completes, we appreciate more what is meant by Whitman's "the body electric." Our unitary processes are no longer generating electricity. There has been a power breakdown or, if you will, an animate blackout. Yet all the electrical apparatus is still there.

(18)

If humans were immortal, we would live even more under the yolk of the past, if there would have been room for us to be born. Truth to tell, life's completion enables progress and all of us alive today owe our very lives to it.

Without it, the earth would have long ago been in the thrall of ancient tyranny and populated beyond its capacity to provide space for new life. What we call "death" actually helps make new life free and possible.

(19)

When our unitary lives complete, every atom that makes us up is still alive and capable of combining into new forms of life. So I do not think it is wise to lock our remains away from nature's processes. Instead, I think we should imitate nature.

When a solar system completes its animate lifespan, the sun flares and incinerates the cooling planets. I myself would much prefer to have my remains oxidized and then sprinkled on a grassy field, so that they are free to rejoin the life cycle as soon as possible. I might request that a small portion of them be spread where my family is buried or perhaps interred there in a small container. I suppose I could have a gravestone among theirs, even though the atoms that constituted my body at the end would have been set free to continue as their eternally vibrant selves.

(20)

Here is a little prayer, or hope, at completion: I have done the best I could with my life and feel privileged to have had a share in life. I have found sufficient joy to be satisfied. I bless those I love and all other living things. I bless the Universe and any other Creator or Creators who are beyond my experience and trust that, if there is an after-reward for good behavior, I will have earned it, because I have taken good care of life, Creation's greatest gift, and the foundation of all human existence. Life bless you and keep you, one and all, now and forever.

(21)

When we complete our individual lives, the sky does not crack open to recognize the event and most of the world goes its way. We live and complete our lives within the locus of our love, friends, and acquaintances. The limited impact is, rather than bewildering indifference, an enabling blessing. What hope would life have if all of us had to participate in the end of each of us? There is, on the other hand, the universal sympathy we are likely to feel for one another's fragility and mortality, and that universality seems appropriate.

(22)

Life invites us to participate in it. Enabled by all of Creation, it is the greatest invitation each of us will ever get. Let us accept it as good, not reject it as unworthy. Let us live in its wonder and be content.

NATURAL DOOMSDAYS

We are so new upon the earth we keep expecting that human life will end, as a supernatural doomsday, and we may wish ourselves thither lest we mend our death-bent behavior. Any natural doomsday is likely millions of years distant, when an asteroid that we cannot deflect strikes the earth or the sun, some 5-billion years hence, flares and wanes into a red dwarf. What is our urgent concern – our own innocent death wish or living on within nature's essentially eternal procedures?

(2)

Since we know life within the biosphere is subjected to mass extinctions – so far separated by relative enormities of time – we can expect that the human race, or a great part of it, has a certain share of the animate span of life on earth. Since other creatures preceded us, we may expect that others will follow us. Our challenge is to appreciate our human share, so we may sanctify it and fulfill its finest potential. The capabilities of life itself and the achievement of life's

Origin in enabling them invite us to a happier destiny that we have yet to grant ourselves.

(3)

We are threatened, moment to moment, with self-inflicted mass extinction. Yet we wonder what we shall do in the enormous by and by when an asteroid pelts the earth or the sun burns out.

(4)

After eons of life, we may indeed face an unpreventable natural doomsday. By then we can rightly see it as the natural completion of human life on earth. But such an eventuality is very likely almost inconceivably distant. Now is our opportunity to secure our natural lifespan. Let us commit ourselves to doing so.

FAITH AND FEAR

Faith In Life contains an element of fear but primarily for a different reason than traditional religions. We know that if sufficient numbers of people do not convert to a belief in the primacy of life, we are far more likely to continue to demean and destroy it and may annihilate it.

RENEWAL

All things remain vibrant through renewal – culture, life, even, it appears, the universe itself.

(2)

How does Creation renew itself? As we now understand, it does so by recycling matter and energy, in the form of atoms and molecules that made up previous unitary lives. The process was called "decay," because we didn't understand the disjunction as the prelude to new conjunctions and that the process is, overall, life-oriented.

(3)

Creation is continuously renewing itself on a universal scale, of which we each are a part; it renews us daily as living individual beings and will renew us entirely, in terms of the matter and energy that make us up, when our individual lives are complete.

(4)

What would the earth be like in its absence of Creation's processes of renewal? Far less savory and very likely uninhabitable. We ought to live in consonance with Creation, so that, as we utilize resources, we do what we can to renew them.

THE DEATH CULTURE

The death culture, which is still pervasive worldwide, is based primarily on the sanctification of death. Its destructive effects can overwhelm us and, at times, make us feel little hope for remediation. The urgent remedy is the establishment of a worldwide life culture, based primarily on the sanctification of life. Its life-nourishing effects can inspire us and grant us abundant joy.

(2)

Wherever humanity goes with its somber rejection of life, instead of the joyful acceptance of it, we will kill it.

(3)

The majority of the human race is still primarily death-directed, particularly in its most sacred beliefs. As a consequence, we are directing ourselves primarily toward our own destruction.

(4)

When we refuse to let life live naturally, we kill it, either its natural behavior or literally. Amid our activity, even our laughter, life as created does not exist – as a gift bestowed by Creation, unencumbered by any inherent detraction but pure as a newborn's innocence, that invites our faith in it and our mutually considerate fulfillment.

(5)

Since most people are still alienated from life – as a gift to embrace, fulfill, and be content with – its potential remains dormant, stillborn every day; in fact, it is perishing moment-to-moment within what serves as our excuse for life.

(6)

Knowing the dynamic of the death culture cautions us to be careful. We can observe in innumerable instances that life creates a tide against death; to resist, death shapes itself into a force against life, which is often murderous. Said another way, and more easily in evidence, the good create a tide against evil, and evil shapes itself into a force against good. Good does not kill but evil does. The historical record speaks for itself. So all those who advocate life should only dare with care. Your first duty is to your own life – and to live as an example of what Faith In Life can inspire.

(7)

Only a primary faith in and reverence for life, instead of a primary faith in and reverence for death, can rescue us from our relentless drift toward individual and collective catastrophe.

(8)

Our conversion from death-orientation to life-orientation is no less than the pivot of human history and the determinate of our destiny.

CULTURAL DEVELOPMENT

All positive cultural development is life enhancing. The denigration of life in any form is an effect of our "sickness unto death."

(2)

An indispensable part of the development of culture is the renewal of religion, which is the fundamental event that can nourish our spirits and give our lives a saving direction.

(3)

The life-culture awaits the creation of its own culture.

HUMAN HISTORY

The overall movement of human history is, I think, from death orientation to life orientation. Our very survival depends on a timely change.

(2)

Until recently, we could take for granted that, as a species, we would likely survive. The assurance no longer graces our lives.

(3)

We face our history as a species and acknowledge that, as so far recorded, it consists of a mixture of spectacular achievement and horrific bloodletting, with the more usual event the latter. A great deal of the bloodletting has been done with cavalier insensitivity to the sanctity of life, enabled by beliefs that subordinate life as transitory and expendable.

Our challenge is to grace our achievements with behavior toward one another that is commensurate with them. How may we accomplish it? We must make the sanctity of this life widely understood and primarily compelling, not only for ethics but for religion. If we are to initiate a history commensurate with our life-oriented potential, we must enlist the spiritual impetus of religion by understanding that revering this life is the most credible and direct way we can lead virtuous lives and express reverence to the Ultimate Source of life, which we choose to regard as God.

(4)

We might see the spiritual dynamic of history in three stages: living primarily for another life with illusionary spiritual stability, much inexplicable sadness and persistent abuse of life; a period of indecision, with much spiritual instability, perplexity and continued abuse of life; then living primarily for this life, with soundly based spiritual stability, a clear commitment to the care of life, and abundant joy.

(5)

One day, let us hope, we may be able to divide our history into the life-rejecting stage and the life-accepting stage.

(6)

In our infancy, we invent fancies; in childhood, we discover they are not true and become disappointed and confused; in adulthood, we understand that the truth is greater than our fancies and regard it as sacred.

(7)

An unprecedented worldwide effort has become urgent to assure that humanity does not negate and destroy the hard-won enhancements with which modern civilization has enabled and blessed life. The churn of civilizations lost, enumerated in woeful detail by Toynbee, is like a series of worn stone markers that point toward a likely sequel and indicate the fragility of our own civilization and the extent of our challenge. We must not allow primitivism in any form to subordinate and devastate our heritage of achievement.

(8)

While it is advantageous to keep up with the news, we also require a larger vision. History just doesn't happen at the rate of 100 pages or 30 minutes of news a day. The current century will one day be summarized in fewer words than a usual day's reportage.

(9)

In a backward county or country, one realizes the old darkness still dominates – life, demeaned, with little, if any, possibility for the populace to find fulfillment, joy, and satisfaction in it.

(10)

Industrial development occurs in a sequence that makes sense: early industrialization, which separates people from nature and pollutes; then the improvement of industrialization, which reunites people with nature and cleans up the environment.

(11)

A system based on enlightened natural behavior would only become possible after a degree of development.

(12)

History's debacles mock even the suggestion that we have been born into a world of great promise. Yet we can see, in the life-oriented achievements we have managed to wring from our early and recent history, that we do have the capacity to enhance life and diminish evil – enough, in fact, to infer that we may work toward the only completely satisfying goal: life-devoted behavior ideally established worldwide.

(13)

It is a mercy that early humans were unaware of human life-oriented potential as time has revealed it. Their scope enabled them to enjoy the fulfillments that were available to them, as an infant can enjoy its relatively basic fulfillments without being troubled by the fulfillments available to adults.

(14)

The past, present, and future are all part of the human epic and, in their life-enhancing achievements, sacred to life.

(15)

What course may the individual take to modify history? While some will attempt extraordinary means, and hopefully peaceful ones, most of us are left with the effect of how we choose to live. If enough of us make the correct choice, history will be changed.

(16)

Out of the present, as moments in succession, emanates the future. Over time, life-enhancing choices can become preponderant in human history and the long darkness subordinated by welcome light.

CHANGE

Many people are so solemnly persuaded that life is a misery to suffer that convincing them it is actually a blessing to enjoy may be one of history's most difficult challenges.

(2)

Let us have Faith In Life and do what we can to enable life-sanctifying change while we preserve ourselves as examples of life conducted in fulfilling and mutually caring ways.

ETHICS

When we understand the greatness of the gift of life, a complete ethic is immediately evident, in fact, a complete religion, based on the care and fulfillment of life.

(2)

Ethical principles that are equal to life enable it, while ethical principles that are not equal to life diminish it. We can find ethical principles that are equal to life in only one way: by looking into life itself. The wisdom inherent in it exceeds our wisdom to invent an alternative. No wonder. Life is Creation's greatest creation.

(3)

Our behavior occurs in the natural world and our principles of behavior ought, logically, to be derived from it. They are not elusive and uncertain; in fact, when we concern ourselves with living and values fit for it, they promptly become self-evident.

(4)

We may guide our lives simply by thinking, I know my life can be, to the extent that it is within my control, what I decide to make it, and I want to make it the best life I can.

(5)

We behave virtuously two ways: by doing things that are good for life and by opposing things that are bad for life. Both kinds of behavior give meaning to and distinguish our lives.

(6)

When we acknowledge the existential concept that we define our lives through our choices, we then ask by what standard we choose. We may turn to Albert Schweitzer's ethic of Reverence for Life and his thought, "What is good for life is good, and what is bad for life is bad."

I prove the validity of these concepts with reference to the greatness of life.

Even if a skeptic maintains they cannot be proved, they work. In fact, they can save us. So they have indispensable pragmatic value.

(7)

A description of "the good" is not complex when our standard is the correct appreciation of life. We may also note extremes of good and evil. Absolute good can be described as life nourished and lived at its life-affirming potential, while absolute evil is life diminished or killed at its life-negating potential.

(8)

We aim for life preserved for its natural span and perfected to the extent it may be, while we oppose life endangered or extinguished before the completion of its natural span and debilitated in any preventable way.

(9)

When people misbehave in a life-demeaning or life-destroying manner, we may rightly see their behavior as a desecration of life, which any life-appreciating society would consider an offense; in worse cases, a crime; or, if you prefer, a sin against life.

(10)

Many things that tradition considers evil are, in fact, good, because they are good for life, but tradition has often grown out of praise for what is good for death. As a result, it is understandable that some aspects of a life fulfilled in its natural proportions have traditionally been demeaned, scoffed at, and considered downright bad.

(11)

All creatures are capable of problem solving. Especially adept in many ways, we take pride and find heroism in it. And what would we do with our capacities to solve problems if all of our problems were solved for us? Then we must ask, why do we have problem-solving capacities? Because we have been equipped with, that is, we have evolved with, the ability to solve our own problems and go on to lead satisfying and distinguished lives.

(12)

The existence of good and evil invites us to choose what is good and, in so doing, help the proportion of good exceed evil.

(13)

When we esteem life, we have a harder time harming and killing it and may actually be incapable of harming or killing what we sufficiently esteem. We make it easier to harm and kill by distancing ourselves from and reducing the value of what we intend to harm or kill, whether animals or other people, as in referring to them as "only animals" or "the enemy." The fact that we must make such an effort is a sign of hope. It means that our natural disposition is to care for and preserve life.

(14)

An inherent part of an ethical commitment to life is a desire, not only to live a distinguished life, but to ease the burdens that poverty, ignorance, governmental abuse, and other forms of debilitation heap on it – to help lift the rocks off the backs of others so that they, too, may experience the joy of life and thereby learn to care for it, too. It is also a part of such an ethical commitment to do what we considerately may to lift, or at least to take a bemused interest in lifting, the blinders on those who do not see the value of life and what its greatness invites, in fact, with its own quiet voice, ever suggests.

(15)

Sometimes, we may experience feelings of helplessness or futility, usually as a result of reflecting on our smallness in the enormity of the universe or when the probabilities of life seem to go tragically awry. At such times, our values can seem to evanesce, and we may suspect we live in an infinitely malleable ethical universe, but as soon as we imaginatively push the borders of behavior, we realize that there are limits. The basic guide that we can always turn to, as certain as sunlight when clouds clear, is the undeniable overall value of life itself.

(16)

We are a bit like sponges. We absorb a certain amount of whatever we are dipped in – in goodness, goodness; evil, evil. We need, in our still-life-negating society, from time to time to wring ourselves out, so that we may experience afresh what is truly valuable and devote our lives to it.

(17)

"The greatest happiness for the greatest number," in Jeremy Bentham's phrase, is a laudable effect but without a credible cause, unless we take the Utilitarian idea that pleasure is good and pain is bad as a credible cause, until we realize the overwhelming value of each life and the person's, or other creature's, inherent right to happiness. Then we know for certain why we should each behave in ways that help effect the happiness of all.

(18)

When we realize that life itself is our fundamental value – in fact, within life as we can assuredly experience it, our highest value – the respect we formerly directed beyond life can now be directed toward it. As a result of this new communion with life, we can experience intellectual and spiritual rewards that are every bit as great, and that rest, we can be sure, on a solid foundation.

(19)

In a succinct paraphrase of Plato, "We become what we do." This is the iron bar of natural justice. When we do good, we become good, and when we do evil, we become evil. The good enhances our lives, and the evil debases them. So, when we value our lives, we know to do good.

(20)

We often call life itself our fundamental value, so that we do not exclude the possibility of value beyond life. But we may also decide, at times, that we are convinced, to the extent we may be, that this life is our entire gift. So we may maintain, at such times, that life itself is our highest value.

(21)

What we value merits our respect, and what we value most, our veneration.

(22)

Schweitzer's ethic of Reverence for Life is a sound everyday guide that also includes implicit reverence for the Origin of life. But I think Reverence Through Life makes the implicit meaning explicit.

The first can clearly guide our daily actions, while the second can clearly satisfy our longing to express our reverence for the Source of life, whether the Universe itself or the Universe and a Creator or Creators outside of our experience of it.

(23)

Despite Dostoevsky's frantic fear that, without supernatural superintendence, "everything is possible," it is not. Inalterable and distinguished guides to conduct are inherent in life itself.

(24)

We have reached a juncture in our evolution when we must commit ourselves to life-sanctifying beliefs and the behavior inspired by them to save ourselves from the torments of self-inflicted catastrophes or annihilation.

(25)

All timeless values are life enhancing, for instance, love and kindness, achievement and excellence. The perception invites us to become timely representatives of such values and thereby actuate them. They may be compared to Plato's Ideas, his universal abstracts of which specific existents are representatives. Unlike, however, the universal idea of a thing, for instance, a chair, which is manifested as specific chairs, timeless values require us to become living embodiments of them. Such values are among the foundational inspirations of a life-affirming religion. They guide us, individually and socially, toward our finest potential.

(26)

We seek the expression of our true being, our unique human nature, unencumbered by the liabilities and perturbations of the death culture. Since we still inhabit it, a special effort is required to see through its life-demeaning assumptions and affirm our natural inclinations. Of course, inclinations, and especially first ones, are not always commendable and we have our life-enhancing judgment to assess which ones are. Over time, our true being, freed to live and think as we are born, toward life, will let us evolve new traditions that are intelligently suited to and wholesome for life.

(27)

Existence as we experience it seems to indicate that what has evolved us has enabled our freedom to choose and has provided time for us to grow in understanding, individually and as a species, so that we may learn to make wiser choices.

(28)

Does not "the silence of the universe" indicate that we are meant to discover the guides we need for our lives on our own and where else but in life itself? Certainly, it does not invite us in a more credible way to assign the task to an authority outside of our experience of life.

(29)

Evidently, Creation has Faith In Life as it would evolve. Life appears to be gifted, as the saying goes, without any strings attached. We are not mere marionettes in a play conducted by outside powers. Fortunately, life is too grand for such a debilitating presumption. Then let us be as generous in our appreciation of life as Creation has been in gifting it and acknowledge that life is a freedom-graced gift, unencumbered except as it must be for animate beings who find themselves as part of the natural unfolding of the universe. Life's care and conduct are in our hands. We should accept our freedom and act on it in ways that are worthy of the gift of life. Thereby we may best distinguish ourselves for ourselves and for the Source of life.

(30)

Contrary to the stern-browed objections of tradition, the most benevolent vision of Creation is actually that life has been created for the enjoyment of its embodiments. Since we have minds and fellow creatures, the intent would obviously include the hope for intelligent and mutually considerate enjoyment.

(31)

It would be illogical for Creation to make our foundational beliefs difficult, since few could understand them. They would, logically, be easy for all to understand, or the hope of a well-behaved society would be, by definition, unachievable.

Let us put aside the puzzling constructions of tradition and look at the beliefs we find commendable, all of which have in common that they are life-preserving and life-enhancing injunctions.

Even traditional beliefs that are life-directed are generally easy to understand. If asked to name the most life-enhancing belief in the Old Testament, at least, the most quoted one, many would say, "Thou shalt not kill." Easy enough to understand. Moving to the New Testament, we find, using the same criteria, what we may consider a positive expression of the same idea: "Love your neighbor as yourself." Also quite transparent, although it does depend somewhat on who your neighbor happens to be. Moving along to more modern times, let's turn to the legal principle of the sanctity of life. Hardly difficult to grasp, either. What about the colloquial and underestimated injunction, "Live and let live"? Easy as can be. So is Schweitzer's Reverence for Life and our Faith in Life, Faith through Life, and Devotion to Life. In fact, all of these life-preserving and life-enhancing beliefs are so easy to understand we're tempted to dismiss them as not entirely adequate to, what we have been convinced, are our complex needs.

(32)

If people are to value life, they must find pleasure in it. So pleasure, considerately obtained, is life enhancing. It is more than to be preferred to pain; it is a foundational virtue.

(33)

Since early human life was often harsh, our ancestors found it hard to appreciate the gifts of life. Yet human life is now far less harsh and, interestingly, just when, to preserve it, we need to reevaluate our estimation of it.

(34)

Natural behavior is good for life, because it lets life live. It also affirms the rightness of what Creation has made. It is therefore twice virtuous. Unnatural behavior is bad for life, because it prohibits life from living. It also denies the rightness of what Creation has made. It is therefore twice bad.

(35)

Life as we experience it is part of Creation's will to live, its devotion to life. When we affirm our will to live and devote ourselves to life, we live in consonance with Creation.

(36)

When we define God as the Ultimate Source of life, we may say that whatever preserves and considerately fulfills what God has made is virtuous and whatever harms or illogically represses life is evil.

(37)

The ultimate achievement of the universe – the vibrant result of its fiery stars and variable planets – is life. So we may say that, above all else, God made life. By God I mean, as I always do, our natural God, the Universe itself, with openness to the possibility of a God or Gods outside of our experience, who evidently have allowed us to function with the resources we have been gifted with.

(38)

The more well-adjusted people are to life, the happier they usually are, while the more maladjusted they are to it, the sadder they often are. The essential difference between them is delight in or denial of the natural joys of life.

(39)

Morals should not be against natural life, but for its considerate enjoyment.

(40)

Moral standards consonant with life make natural behavior seem virtuous, but moral standards inconsistent with life can make natural behavior seem sinful.

(41)

Each time our behavior expresses Reverence for Life, we enhance our lives, all of life, and express reverence to the Originator of life, while each time our behavior expresses irreverence for life, we degrade our own lives, all of life, and express irreverence to the Originator of life.

(42)

Mutual consideration is an expression of Reverence for Life. So, in a life-oriented society, it is a moral imperative. Conversely, living without mutual consideration is an expression of irreverence for life. It diminishes one's own life and the lives of others.

(43)

A life-affirming society is "clean," because its morality invites its citizens to express their natural life-oriented desires; that is, it allows such desires to live. A life-negating society is "dirty," because its morality requires its citizens to repress many of their natural life-oriented desires; that is, it kills such desires and offers in their place unsatisfying, degrading or often violent substitutes.

(44)

We can say that a virtuous life is one that if founded on the wisdom and the will to live an intelligently fulfilling and mutually considerate life.

(45)

I have made many mistakes. I am fallible. Realizing I am is part of what enables my efforts; I know I will be successful a certain proportion of times. Being fallible is also part of what makes me human and proudly so. I am free to strive and to do my best.

Most deeply, I have had a great and guiding feeling of sanctity about life since my late twenties, when I first realized how worthy of worship life is. I have always tried, especially since then, to do my best, never intentionally hurt anyone, and help others when I have been able to.

Each day when I assess myself, and at the time I reach the completion of my life, I hope to say, and think I will be able to, that I have tried to do the best I can with my life. I think that is what being worthy of the gift of life requires and no more, especially by the Source of wisdom as great as that which is manifested in life.

(46)

What each of us decides to make of his or her life becomes part of what we will all make of it.

(47)

Life's deepest joys spring from life-directed behavior.

(48)

We accept beliefs that enable and enhance life but reject those that disable or diminish it.

(49)

We should always behave in a way that is worthy of life.

(50)

When we behave as we wish the human race to, our portion of the human race does.

(51)

We have within the appreciation of life enough answers, awaiting enough actions.

(52)

Live primarily, or exclusively, for this life. It needs you to and so does its Creator.

(53)

Live as if your life is part of God's life. It is.

A NATURAL MORAL SENSE

I believe we have a natural moral sense that is based on life itself, but its strength depends on how much we respect life. The more we appreciate the greatness of life, the more we will respect it, so the more we will strive to preserve it and fulfill its finest possibilities.

Even weakened as our natural moral sense often is, we usually feel joyful when we do something life enhancing and distressed when we do something life diminishing. These feelings may have evolved as part of our sense of survival and quest for happiness.

CONSCIENCE

We are our own best conscience, not an overly internalized or a remote substitute. How will we do? We will enable ourselves to be our best, without internalized or externalized guilt. All we require as a guide is knowing the value of life and acting with consideration for our own well-being and the well-being of all other life.

(2)

All of our thoughts and acts are observed by us as individuals, which is quite sufficient, except for those who are alienated from responsibility, namely criminals and the insane. Since we are each part of the universe, taken as our natural God, and self-observant, we may, for additional inspiration, note that we are, in the foregoing sense, observed by God.

RESPECT FOR HUMAN NATURE

When we have respect for Creation, we have respect for what it has created. It has created human life, and the essence of it is human nature. So, rather than lamenting "human nature," which is a form of irreverence for life, we should respect it – and ask what we have done to misshape it. Logically, it should be life respecting, but it is malleable. If our nurture would allow it to develop in a naturally life-respecting manner, we would learn to admire and trust its promptings.

(2)

Since human nature is the nature of beings evolved by Creation, its naturally life-oriented promptings are logically wiser than rules of conduct arbitrarily invented by humans, and we should hearken to them; that is, if we're willing to admit that, at any given stage, we're not wiser than Creation.

(3)

Why would we look down on "human nature" when it is part of the nature of what has been created? It is, as is all of life, an aspect of the divine. Ours is not to denigrate human nature, which is an instance of irreverence, but to create a world that nourishes human nature toward the actuation of its finest potential.

(4)

The promptings of human nature contain, as an aspect of our original natures and of our development, an appreciation of, not only individual value, but mutual value. We understand very early that we are more likely to survive and enjoy life united, instead of divided. But we evolved our practical understanding of mutual benefit at a time when we viewed it in terms of clans and tribes. We need to enlarge our perception to encompass the vision that we are all members of a global community, as well as one species among all the other species.

THE HUMAN SPIRIT

We should never underestimate the vitality and grandeur of the human spirit; it is the spirit of Creation within us.

(2)

How much subordination and abuse of the human spirit can humanity endure before it demands a reaffirmation of the congenial values that are fundamental to a civilized society?

(3)

It is good that the downtrodden do not forever suffer injustice. Their resistance is a sign of the vitality of the human spirit and its will to live. It calls attention to the necessity of rightful remedies.

A NATURAL WILL

We do not have complete control over our lives. We live within a multitude of givens, which results in probability. But we do have, I think, a logical degree of decision-making capability, which we may call, not free will, but a natural will. It provides us with sufficient freedom to determine the overall character of our lives, particularly barring unusual duress, usually inflicted by our fellows. When we exert the amount of freedom we do have responsibly, we do the most we can to lead ethical and distinguished lives.

(2)

If the creatures of Creation, particularly its most intelligent ones, were to evolve with a natural degree of volition, the result would be more distinguished than if they were to evolve with no volition. The way life has been created manifests a wisdom that would, I think, prefer the more distinguished choice.

(3)

Yes, the universe is enlarging toward its own overall destiny, but even if the destiny of the whole in which we exist is determined, the universe enables within its span, I think, an appropriate degree of volition for its animate creations, especially its most mentally developed ones.

(4)

If life is logical, and we believe it has evolved to be, we have the quantity of will we need to conduct our lives in responsible, distinguished ways and, if Creation is benevolent, and we believe it is, to conduct them in ways that are rewarding. The great question of will is, not whether we have free will, but what shall we, with our quantity of natural will, make our lives?

(5)

We have evidently been given a natural amount of independence. The first step is to acknowledge it. Then we can vow to make ourselves worthy of it.

(6)

Life can be as good as we can make it. The possibility may be considered the charter of our natural wills.

MENTAL LIFE

The fulfillment of our finest mental potential is a foundational aspect of a religious commitment to life.

(2)

Creating consciousness is what the brain does. All thinking, awareness, and other physical activities are emanations of and feedback to this astonishingly capable natural organ, our, if you will, animate central processor.

(3)

The mind is like a prism, dividing and expressing the light of mental energy in a rainbow of vibrant colors, which we may wander among with wonder. I do not mean, of course, in a hallucinatory manner but am speaking metaphorically about the way the brain divides mental energy into an array of uses, based on voluntary stimuli, such as our desire to cogitate, and involuntary activities, including the conduct of its own vital functions and those of the rest of the body.

(4)

The pursuit of knowledge, wisdom, and creativity is, generously, not only distinguished but, in Keats' words, "a joy forever."

(5)

At any moment, we can be conscious of ourselves as totally alive – a thinking, breathing, feeling being, who has at this moment the privilege of being part of the life and awareness of Creation.

(6)

Since the human mind is part of the way Creation is conscious of itself, we can, by the ways we choose to be conscious of Creation, distinguish ourselves as worthy of it.

(7)

When we have a talent, we have an obligation to the quality of our lives to manage it so that it becomes a blessing, not a liability; a source of joy, not of agony.

(8)

In an erect species, the brain is located at the top of the body. But Aristotle's mind-body question, in terms of which is superior, is biased. We can get along without parts of the body, but what are we to do without the mind, which administers our entire being?

In addition, at different times and for different people, either the mind or the body is more important than the other.

So, when Aristotle asks what we would rather have cut off – our heads or our sex organs, we can only smile, as we do when we read his works on natural phenomenon, which express the errors of his time, and so we turn to his ethics and metaphysics, which express his wisdom.

(9)

The mind appreciates the body's pleasures, and the body, the mind's. Achievement by either delights both.

(10)

If you believe life is without merit, you believe you are smarter than Creation, which decides every moment that life is worthwhile. As Montaigne says, "Man imagines gods by the dozen but cannot even make a worm."

(11)

I believe in the unity of each life. The mind and the spirit are the ongoing creations of the vibrant body – the mind of the brain, the spirit by the brain and its extension, the nerves, especially in the conjunction of the solar plexus. They are manifestations of the body's life energy, natural glories of it, as are the senses, which are, in fact, the brain's external receptors, which tell us about the outside world reliably or their functions would be contradictory.

(12)

We have an overall spirit of life, of which thinking, physical activity, and feeling are the general expressions.

(13)

We have enough intelligence to guide our lives or life would be a contradiction in design.

(14)

Everything in nature seems to have intelligence, even subatomic particles, which combine into different atoms, a feat that still partly defies even our most advanced understanding. We may say, then, that intelligence, at various levels, seems to be universally present, and our own mental lives are a gifted part of it.

(15)

While our minds have more capacities than most of us can hope to measure, we can do our best to fulfill them. When we understand the greatness of the mind, we're more likely to want to fulfill its finest capacities, as part of our duty to ourselves, life, and its Originator or Originators.

(16)

The basic processes of nature are perfectly rational, from water tumbling down a brook to the conduct of birds. Therefore, we can infer that irrational behavior is, in a healthy and peaceful mind, unnatural. As a result, we should encourage people to behave naturally, which, one expects, would be rationally.

(17)

What are the goals of mental life? Voluntarily, the fulfillment of the body's potential, including that of the brain. Involuntarily, the conduct of the ongoing functioning of the body.

(18)

The mind presents a process; we and the world we inhabit provide the content. Let us choose wisely and hope for a world that provides life-enhancing content.

(19)

Every animate being contemplates at its own level and enjoys the process. The tool making of chimpanzees and even the eyes of pets are conviction enough.

(20)

The conduct of our mental lives is part of the care of our lives and is therefore part of a life-devoted religion. So let us have great faith in our mental lives and devotion to their fundamental and most inspired possibilities.

TRUTH

Truth is a light from antiquity to the present, shining in the darkness, like a candle or a spotlight, helping those who love the truth see their way.

(2)

Truth is not polymorphous but singular – a unity to be treasured and discovered.

(3)

Believing something is true, no matter how intensely, doesn't make it so. Reality does not accommodate itself to our hopes; it is for us to accommodate our hopes to reality.

(4)

An intricacy understood enhances our lives more than a mystery awed.

(5)

There is more truth in unassuming wonder than in unsupported answers.

(6)

We finally realize that any trust and even truth itself may be traduced. We can only hope they obtain against undue influence.

(7)

Our vistas in the natural universe encompass at least aspirations toward the infinite; so a philosophy that is equal to our inclinations must encompass, not only a view of the natural universe and our conduct in it, but a credible consideration of our disposition toward the infinite.

(8)

To paraphrase Locke, while we cannot know everything, we can know enough to guide our lives and to understand our place with regard to what we cannot know. We can add our own point of view that, if Creation is logical, what we cannot know we need not know.

(9)

Truth comes slowly but time is young.

LOGIC

The logic is direct. We revere what we value. We value life. Therefore, we revere it. More religiously, we reverence what we most esteem. We most esteem life. Therefore, we reverence it. As a final instance: We worship the Ultimate Source of Life. The Universe itself or a Creative Power beyond our experience is the Ultimate Source of life, and we include both in our concept of Creation; therefore, we worship Creation.

(2)

Every argument consists, as you know, of a thesis and an antithesis. We ask if there is a third answer, or a synthesis. To arrive at one, we can, very simply, triangulate the two opposite points by drawing a triangle up from them. At the peak, we place the primary thing both sides have in common. Next down, the second idea they share. When we get to a level where the sides agree on no more ideas, we can go no further to reconcile them.

But now we are in a better position to ask if there can be a third answer, or synthesis, which will create an upside-down triangle on top of the peak of the first triangle. Soon, in thinking about the new answer, the human mind will develop agreement with it, perhaps modified, and disagreement. These two points form the counterpoint of the new dispute and the top of the new triangle, which also serves as the base of a newest triangle.

How may we apply this visualization? There is an overall question that pertains to all disputes – and is the most important one: the dispute cannot bear on any entity that is not within the knowledge of the disputants. The most urgent application of this understanding is to the thesis of faith and the antithesis of science. Whether or not there is a Creator or are Creators outside of our experience cannot be determined within the point-counterpoint of the dispute or the triangulation of it. The entity remains outside of the reach of the triangle.

Religion can maintain that such a Creator or Creators must necessarily exist but cannot prove the point. Science can only say that we see no evidence of unnatural influence and intuit what we may from that. But the actual subject of the dispute remains beyond the locus of the disputants' knowledge and therefore beyond the contents of their argument. As the saying goes, "We can't get there from here."

Accepting the limits of true argument only disproves or must leave unknowable what is obviously beyond the locus of our knowledge.

WISDOM

True wisdom sees the value of life and needs to know no more.

(2)

Wisdom allows the balanced enjoyment of life, which can, at times, be exultant.

(3)

If everyone decided to care for and enjoy life prudently, who would be left to demean and destroy it?

KNOWLEDGE

We do not need to know everything to know how to live and cannot, on the level in the universe that we evolved, be expected to know everything. But, if the unfolding of the universe and the evolution of its animate life is logical, we can know what we have to.

(2)

Our unattainable need to know, or the pretense of knowing, everything in and out of the universe bespeaks an ill adjustment to and ingratitude for our evident level of being, as well as dissatisfaction with or ignorance of all we can rightly know.

(3)

The quest to understand life and its source is an inevitable aspect of the development of intelligent beings in a naturally evolving universe. Our efforts would necessarily be based on our capacities at the various times we make the efforts. So our earliest efforts would be founded less on true knowledge and more on mental faculties, slightly informed, such as imagination, intuition, and reason. As our knowledge increases, our efforts can be founded more on it. For instance, we now know we can arrive at true knowledge of the physical world though the experimental method, and we can make other intellectual efforts that are supported by informed reason, imagination, and intuition.

(4)

Our earliest ideas occurred when we were at the level of the clan and the tribe. They were usually expressed by the leader or shaman and, while they were inspired by our mental capacities, they could not be based on extensive knowledge of the natural world. Since what comes out of the mind is necessarily dependent on what goes into it, the ideas were more likely to be erroneous. As our knowledge has become more widespread, our ideas have been expressed and judged by a more diverse and informed group of citizens, and, as our knowledge has increased, our ideas have been more likely to approach the truth. While our earliest ideas still comfort and enable many, our later ones can often do so more credibly.

(5)

We know more than any other creature on earth. Yet, caught between the basic knowledge most creatures have and the intelligence to learn more, we feel we must know everything, even if we must invent what we cannot know. So we are often caught in between what we can know and what we unrealistically strive to know, casting ourselves as a sort of strident "monkey in the middle." We ought to accommodate ourselves to the enlightened realization that we can have what we might call sufficient knowledge.

(6)

All the knowledge we can have is within life, and life itself is the greatest thing we can know. Let us live within it with appreciation, gratitude, trust, and devotion to it.

(7)

When we have Faith In Life, there are unanswerable questions we no longer have to puzzle over. While we may still enjoy speculation and investigation, we are comforted by the knowledge that we believe in the greatest thing we can know and have faith in it.

(8)

Our fundamental knowledge is based on our understanding of how the universe works, including the beings it evolves. The universe contains, obviously, more knowledge than we have yet discovered.

(9)

The universe tends toward and expresses its highest achievement in animate life, particularly exceptionally intelligent animate life, which can function as the individual but widespread self-awareness and judicious activity of the universe. Let us be guided by knowledge that is greater than our own and take life as our own highest value. Let us live in consonance with the will to live of the universe. Let us care for life and fulfill our own life-directed potential. To imagine a greater calling is to propose to outthink the universe.

(10)

The most distinguished achievement is to express, not a small idea complexly, but a big idea simply. Unnecessary complexity, particularly in nomenclature, may enable feelings of distinction, ever specious, but are an obstacle to enabling the universality of knowledge. An idea itself is what is precious. The understanding of it is what is distinguishing, and the clear communication of it a necessity. By this approach to knowledge, our separate ideas may become part of our universal knowledge and the light whereby we may illuminate our common way forward. The wisest goal, admittedly ideal, is to make all knowledge available to all.

(11)

We speculate best when we remember what we know and look up what we don't.

(12)

We should never be upset when people disagree with us but expect disagreement and know it is normal. As soon as one human expresses a point, another human is bound to express a counterpoint. Such is the nature of human activity, either in the pursuit of consensus or merely perversely. When a point can be made without eliciting a counterpoint, we must often suspect the workings of authority.

(13)

Not only know but value thyself; in fact, consider thyself sacred.

(14)

I decided when I was in my first year of graduate school, to learn, not only one subject well, but every subject generally. I did not want to have a mindscape that was made up of one tall tree with saplings all around. I believed a broad foundation would be necessary for speculation that would do justice to our current knowledge.

I was living in Philadelphia at the time and, to effect the plan, I went to the bookstore at The University of Pennsylvania. I found the booklet that presented the syllabus for every course and bought, over time, the basic textbook on every subject I felt was important to know the rudiments of. Reading the books as a self-assignment was a particular pleasure, because I was only doing so for myself and my goal was to learn, except as the details attracted my interest, the basic content. The process changed my mental life and gave me more confidence in my contemplations. I recommend it to anyone who would like to view and contemplate existence with a broad foundation.

(15)

We sometimes hear the lament that our knowledge has grown so much that attaining an informed view of our situation is difficult, if not impossible. Difficult it is, but hardly impossible. While detailed knowledge is overwhelming, essential knowledge is learnable. In fact, the knowledge on which a comprehensive outlook can be based is not so enormous that an interested person cannot, with enough diligence and time, acquire it. The acquisition of at least a general overview can be facilitated by books and other media that present overall knowledge or a vision based on it.

(16)

Since we experience the world as a whole, we can best guide our lives and achieve our best when our knowledge is, to the extent we can unify it, whole. The world is one and our knowledge of it should be, too.

(17)

We ought not to mistake the daily information with which we are inundated for knowledge. It is transient data that is more appropriately only an aspect of our everyday busyness.

(18)

I believe Creation is reasonable, as evidenced by the balance and natural functioning of its parts. So I believe that, to a degree that is necessary for accuracy, we can gain accurate knowledge of the outside world through our senses. I believe, in fact, that stimulus essentially equals perception – as, for example, a bright or dim light may be perceived as such. We perceive outside stimuli; therefore, we can know what is outside of ourselves. As for people who do not believe in cause and effect, but only regular sequence, I suppose they have never played sports or games like billiards.

(19)

The more we learn about natural Creation, the more we will know and can infer what might logically be attributes of its Origin.

(20)

I would rather know a little truth than be confident of a great falsehood.

(21)

While we have learned much, we still appreciate little.

PHILOSOPHY

A credible philosophy of life grows out of the understanding and appreciation of it as the natural biological miracle it is, in a universe that has evolved it and continues to sustain it. Such an approach encourages us to develop a sacred feeling toward life. Then all of our beliefs become apparent. We ought to take good care of life and fulfill its finest potential. As a result, we will lead the finest lives we can and, if such a thing is possible and the universe is logical, we will by so living please whatever originated life.

(2)

We can only have a coherent belief system for life when we appreciate life; indeed, until then we hardly merit one, because, if we do not appreciate life, where is our deserving? On the other hand, when we appreciate life, a coherent belief system is readily apparent.

(3)

Only life itself is equal to life; therefore, only a philosophy based on life itself is equal to life.
We know with certainty that we are alive. What should we do?

(4)

I prefer to reason from the earth up, rather from the sky down. I begin by wondering about what is right here before us, life, and only then might I allow myself the liberty to speculate about what might be beyond it.

(5)

Life is credibly our *summum bonum.* Without it, what have we? We would not even be able to contemplate other values. When we realize it is the foundation of all we experience, we can naturally value it in accordance with its merits. And what does traditional religion promise, but an infinite continuation of life?

(6)

Humanism is commendable but our concern must be all of life. Naturalism excludes superstition but also our propensity for infinite concerns. I prefer, as I've noted elsewhere, the name Lifeism, and I'd be comfortable being called a Lifeist. Life Itself or simply Life often works as commensurate nomenclature, too.

(7)

I find Albert Schweitzer's ethic of Reverence For Life foundational. I also think what he considered the "mystical" aspect of it can be expressed as Reverence Through Life, as our most direct and credible worship of whatever has evolved life.

(8)

Spinoza calls infinite extension a necessary perfection of an all-powerful God, but cannot self-limitation also be considered a necessary attribute? And does not the latter more likely allow other perfections of such a Being, among them a self-defined shape? Then might there be a border to our Pantheistic concept of God, beyond which there is nothing, at least, until the universe expands into it, and perhaps an edge beyond which there is simply nothing and therefore no admittance?

(9)

Heidegger states that Being toward its end is our natural state of mind. But Being toward life is the concept that is consistent with the existence and potential of life. More exactly, the concept that is consonant with the will to live expressed by the universe is Being in life with Being toward the realization of its life-oriented potential – a natural dynamic that requires us to settle comfortably into life and that can only be enabled when we do not continually debilitate our lives by emphasizing that they will end, a liability that is irrational. When we live toward life's end, we are ever half in and half out of life, an illogical state that creates stress and obviates our natural participation in and fulfillment of life. Let us embrace our natural reality: Being in life.

(10)

Augustine, tangled in erroneous assumptions, says, "God foresaw all things." Yet, he says, we are still free, because we do what God foresaw in a free way. The poor soul was trying to reconcile the omnipotence of God with moral responsibility. Then he proceeds to blame the so-called Fall on man. He could not realize that, through evolution, we rise to life and may enjoy its blessings.

His small base of knowledge inflicted other complications on his thinking. If God foresaw how we would act, in Augustine's view, badly, then God was a bad, as well as an incompetent, Creator. If we follow Augustine's line of reasoning, we must arrive at the conclusion that the "Fall" was ultimately God's fault. Thankfully, since we now realize that our arrival here is good, we can redeem his God for letting us "Fall."

As noted above, we experienced a "Rise" to life, from the basic ingredients of Creation, matter and energy. We have a natural amount of freedom. And, if any God foresaw all that we would do, He, She, They, or It obviously knew that, with our freedom to choose, we would inevitably do a mixture of good and bad, hopefully, with the accent on good. And we have every impetus to do good. It benefits us, as well as reflects well on Creation.

If a God did foresee everything, we can only hope such a farseeing soul is not bored to tears watching everything work out exactly as foreseen, even unto the smallest detail. Also, if everything was foreseen, we might ask why one would go to the trouble of enabling it? Accordingly, we hope there are some limits to the far-sightedness of Divinity.

(11)

Anselm says, in his "Moralogium On The Being of God," that "there is a Being that is supremely good, since all goods are good through a single Being, which is good through itself." I really do not see why the Universe itself cannot be the supreme cause of all this good. Since we cannot be certain, we have our concept of Open Pantheism to allow for a God or Gods beyond our experience.

(12)

We owe much to those people of reason who brought down the house of unreason, such as Epicurus, Lucretius, Voltaire, Hume, and Russell.

(13)

Rightly understood, philosophy is informed reason, practiced, and religion, truth, sanctified.

(14)

My view of the Ultimate may be described as pragmatic monism with conditional dualism – a belief in life itself first, the monad, and through its service, the expression of our most credible reverence to its Ultimate Source, which is the universe itself as part of the monad, or a Being or Beings beyond our experience of it, which is the conditional duality, in that it may or may not exist.

(15)

We define the Ultimate Source of life so that it must exist, either as the Universe itself or as a Creator or Creators beyond our experience. To have reverence for either or both we need not know the answer.

(16)

The universe is involved in our lives, since we are living embodiments of it, not only evolved by it but continuously supported by it. There is a reciprocal relationship between us and the universe, made up of many facets, from our basic being to our appreciation of and activities amid the life and materials of the universe.

(17)

We know that the universe is involved in our lives and, to the extent that we influence the quality of life in it, we affect its life. The way we affect its life can be an expression of our reverence for it. Whether or not the universe, along with any Creator or Creators who may be beyond our experience, are consciously appreciative of our reverence is immaterial. When we show respect to our greatest benefactor, we do it because we feel we are doing what is due. Such respect is not dependent on whether or not we receive a literal thank you.

(18)

The greatest joy we can know grows out of the fulfillment of our natural life potential – the appropriately proportioned expression of the life energy generated by the ongoing gift of sustenance we call our metabolic processes.

(19)

We believe in pleasure with measure and know that excess, or pleasure without measure, becomes pain.

(20)

Our finite conduct is the most credible way to address our infinite hopes.

(21)

We need to go beyond humanism to a philosophy that embraces our natural fellowship with all life. I've elsewhere suggested the simple term Lifeism.

(22)

It is time for a comprehensive philosophy built on life itself. I trust, despite any faults others may detect in my work, I have made a contribution.

EXAMPLE

What many a philosopher has said we can restate it in our own way: the most influential and exemplary life is one that is lived in a way that distinguishes life and all of Creation. We might also apply the concept beyond our individual lives to an exemplary nation.

SCIENCE

Science, applied broadly as belief in demonstrable truth, has been curative of superstition and prepared the way for a religion based on life as we have come to understand it.

SCIENCE AND RELIGION

Science does not negate religion but false belief. Its discovery of the intricacy and wonder of Creation provides a revelatory basis for a religion based on the truth of existence.

(2)

Let us go beyond the supposed opposition of science and religion and advocate their complimentary roles: science discovers the greatness of life and ministers to its physical needs, while religion sanctifies it and ministers to its spiritual needs.

(3)

My own goal is to observe life as we know it and to do my best to help us evolve a credible religion based on the appreciation of it.

SCIENCE AND ART

Scientists, concerned with facts, have developed nomenclature that is merely functional when it ought to be beautifully expressive and worthy of their discoveries about "nature read backward," the magnificent processes of life and the universe. In time, verbal artists will create apt appellation; visual artists will "illuminate" the word; and musical artists will express it in tone. As the scientific "word" is graced by beautiful language, painting and sculpture, instrumental and vocal music, truth, majestic as life itself, will be inspiringly rendered.

(2)

Contemporary science has been the great benefactor of modern life. We await the benefactions of contemporary art.

(3)

We must judge with care the contemporary nomenclature of science, which is in its infancy. Have we misnamed what we call energy? Isn't it the life energy or basic life of the universe, of which we all partake?

(4)

The enhancement of science and technology with the contributions of art is part of humanity's capacity to be worthy of life and to fulfill its promise.

SCIENCE & TECHNOLOGY

Science and technology enable us to go beyond the struggle for survival to the quest for fulfillment. They do not enable barren lives but make possible lives more secure and enjoyable than any we have known.

(2)

Science and technology not employed for life-oriented purposes will become part of our early, death-oriented past.

(3)

Ours is as much to understand as to invent.

WHAT MAKES LIFE INTERESTING?

Life can be entrancing in its possibilities but also confoundedly complex. Both aspects make it interesting.

(2)

Life seldom surprises us in regard to what happens but how it happens. For instance, we know, generally, we will grow up, go to work, get married, have children and grandchildren, grow old and then our lives will complete. But we usually don't have a clue how these events will happen. In other words, the more general our viewpoint, the more ordinary life seems, the more specific, the more interesting. This same process applies to every aspect of life, from how we view ourselves as individuals to how we view our place in the universe. The delight is in the details.

(3)

We could make a litany of what makes life interesting – overall, the lives of our minds, bodies, and spirits, including our interactions with other people, all the rest of life, and with Creation.

THINKING BIG

When we think big, we can, of course, have big ideas. But, in the sense I'm referring to, we can remove ourselves from concern about the particulars that give our lives their value; our biggest thinking can bring us, more wisely, back to their ultimate importance.

(2)

Space-time is such an enormous value it helps us realize that our lives have their greatest significance to us and to those who are close to us; then to the general welfare of the human race, to all other life, and finally to the great beyond, including whatever "perhaps" our hopes may include.

THE INFINITE

While it is natural for us to want to relate to what we know we are part of, we are first part of life and natural Creation. So our primary devotion should be to life and natural Creation; then, if we wish, our secondary devotion can be to the infinite, whether or not it may exist and whatever it may or may not hold.

LIVING IN FINITUDE

When we acknowledge that we live in finitude, we enhance our ability to appreciate the unique value of life and of all Creation.

(2)

We live in a finitude of natural wonders, or, if you will, natural miracles, yet many prefer to believe in our relatively paltry invented "miracles," mainly because they offer token reassurance of their own life after their time is this often-demeaned life concludes.

(3)

The most urgent achievement for the human race is to acknowledge that we live in finitude, so that we turn our eyes toward life, appreciate its unique value, and care for it accordingly.

(4)

We have recorded our experiences and evolved our thinking enough to realize that we can learn about existence but are unlikely to discover a complete answer to the what, how, and especially the why of it. We live within the universe and all our verifiable knowledge is of the universe. Can we be at peace without what we consider ultimate answers? Yes, when we trust in the wisdom and greatness of life.

Some may feel the adjustment requires a certain amount of resignation. But it also releases us from otherworldly concerns, at least, an excess of them, and we are brought back, in a naturally redemptive way, to this life.

(5)

Do we need ultimate answers to know how to live? Since we seem to be encouraged to live without them, we must suppose that we do not need them, or life would be illogical. Then how do we find our answers? By appreciating all that we have come to understand life to be and vowing to take care of it. In fact, the appreciation of life can be seen as our ultimate answer. It invites us to care for it, fulfill its finest possibilities, and express our reverence through it. When we do, we not only lead the finest lives we can, we show our most direct reverence to whatever the Ultimate Source of life may be.

(6)

What we think or do not think about the infinite does not influence whatever is or isn't. But our behavior in life very much influences our lives and, I think, any logical hope for another life, which we would most logically merit by first taking good care of this life.

(7)

When we accept life as a sufficient gift, we can live in contentment and occasionally muse, if we wish, about what, if anything, may be beyond it.

(8)

We live in finitude, and all connections between what we can know and what may be beyond it are by definition spurious. We can't get from here to there; there is simply no credible connection. We do not even know if such an elsewhere exists. Nor can wishing something were, or were not, make it so. Whatever exists, exists. Our role is to adjust to life as we find ourselves privileged to enjoy a share in it.

(9)

To dwell excessively on what may be beyond this world, imaginary or material, does not seem consonant with the acceptance of life, which invites us to turn our eyes toward it and commit ourselves to it in ways that are worthy of its greatness.

(10)

What if, every time something went wrong, an omnipotent power would step in and right it? Would we be happier? Or would we be inclined to be become rather pitiable dependents? What use would we have for our own moral sense and capacity to right things? Do not we actually have happier lives by realizing we have been empowered to manage and improve our own lives? Is there not ample reason to feel more distinguished by and to bless our apparent independence?

(11)

We live, love, and achieve in finitude. So let us, at least, put our finite lives first. Then, if we wish, our infinite ruminations can have, in logical proportion, a secondary place.

(12)

Actually, we are, if it exists, even more at one with the infinite, in our finite concerns, as we are when our concerns are primarily for the infinite, because we can most directly experience a communion with the Source of life through life itself. So we are closer to any infinite existent when we concentrate primarily on the finite.

(13)

While we have no verifiable information about what may be the transcendent source of the universe, we can intuit backward from what we can know. But we can never be certain of our conclusions. The acknowledgement doesn't debilitate us. It debilitates arrogant enmity based on our different interpretations of what may or may not be outside of our experience.

(14)

There is, in our experience, no knowledge outside of life – neither during it, before it and, we expect, not after it. The only knowledge we can have is that which is available to us in life, and it is on this knowledge that we ought to base all of our beliefs.

(15)

I don't know about what may or may not be outside of the world I can experience. But I do know that, when we appreciate the natural miracles of life and the universe, we can validly believe that we live within the natural Godliness of Creation.

THE SILENCE OF THE UNIVERSE

I see the so-called silence of the universe, not as disheartening emptiness, but as a consistent indication of confidence in life as it evolves.

(2)

When a mechanism is built well, it requires little maintenance; the better, the less; if perfectly, none.

THE INDIFFERENCE OF THE UNIVERSE

Creation may, justifiably, seem and actually be indifferent to the joy we bless ourselves with or the sorrow we curse ourselves with. It may have decided it has provided us with enough to fend for ourselves and to define our own destiny. In fact, competently created life would be expected to make do on its own and to realize its finest possibilities, and we can assume, if we have respect for the competence of Creation, that we have evolved with all the competence we need for the distinguished conduct of our lives. It also seems logical that a creation as enormous as the universe would rely on local competence.

(2)

Many lament that even as we "die" the world goes on without taking any or much notice. But the truth is far wiser. Those who should notice do: the people whose lives we have been part of. But life has evolved in a way that allows it to be a relatively stable experience.

An aspect of its stability is that we are not aware of the completion, or the initiation, of every life. Imagine how ceaselessly distressed or elated we would all be if we were aware of all such events. Finally, would a considerate person desire that the entire world come to a grieving halt at the moment his or her life completed or would he or she prefer that the world go on in the enjoyment of its relative stability?

(3)

When the life of each of us concludes, we can rest content that the people who are closest to us will grieve and, hopefully, celebrate our lives. So, instead of being saddened by the seeming indifference of the world at large, we ought instead to recognize the wisdom with which life has been gifted and be grateful that we have had the privilege to live. Even if our lives conclude when we are alone, we can, by appreciating all that our own lives have been, close our eyes with satisfaction, as well as with the wish that all other life will continue to enjoy its share of life.

BEREFT IN THE UNIVERSE

The perception that we are alone in the universe, in terms of a supernatural overseer, should not precipitate grief, but instead pride, because the intimation is that we have the freedom to choose the lives we will and the competence to chose ones that are meritorious, joyful, and satisfying.

(2)

We are not separated from the infinite but, if it exists, part of it, or, at least, as much a part of it as we are meant to be and likely ever will be.

THE COMMUNITY OF THE UNIVERSE

We may, I think, infer membership in a vast community of the universe.

Why? Wherever animate life evolves, it would likely find itself in similar circumstances – on a planet, or perhaps a moon, warmed by a star, teeming with life, with silence beyond. The environment, as on the earth, would provide its various forms of life, carbon-based on not, the opportunity to live and to realize their potential and provide its most intelligent forms with the ability to improve continually the conditions of life as it can be gifted in a naturally evolving universe.

So intelligent life on each planet must develop a similar history, differing, I suppose, in relatively minor ways, such as language and dress. This similarity would be, of course, more likely true of carbon-based life, especially on similarly sized planets.

(2)

Once we realize that the universe – in its life-evolving elements and life-supporting conditions – provides for the replication of a basic system of life development and support, we can infer much about the evolution and history of the general community of the universe. We need not see, know, or visit other planets to intuit quite a lot.

(3)

We are not alone and hardly need feel that way. What awaits us and future generations is likely quite a get together.

(4)

Since all the matter and energy in the universe resulted from the unfolding of originally highly compacted matter and energy, the entire universe may be seen, in terms of its ability to grow life, as one piece of soil.

(5)

We are, every one of us, throughout the span of our lives, not only inhabitants of the earth, but citizens of the burgeoning universe. We are not alone among the life-evolving sun systems but members of the great and various community of our awe-inspiring universe.

(6)

Since matter and energy demonstrate a great urge to live, we can assume that animate life has evolved everywhere in the universe where the conditions are right for it. So we can assume, I think, that we are part of the great community of life in the universe and will one day be part of it literally.

(7)

I assume that one day people from different solar systems and galaxies will join hands across the universe, sometimes in person and, in more distant ways, through communication.

PLEASURE AND PAIN

Since Creation seeks to live, we assume it provides justification for us to live and help it live. Since pleasure provides such justification but pain does not, we assume life was made primarily for pleasure. Since we have also been gifted with intelligence, we assume for intelligent pleasure. Since we can only achieve our potential individually and cooperatively, we assume for mutually considerate pleasure. And, since we have decided to continue our lives, individually and as a species, we have evidently decided that the pleasure of life outweighs the pain.

(2)

Was life made for pleasure or pain? If for pleasure, we may say, in the Greek manner, the gods are good; for we consider pleasure good. If for pain, we may say the gods are evil, because we consider pain, other than that which alerts us to a physical problem, evil, particularly beyond the extent it mildly sets pleasure in relief.

Said another way, the good will generally will good and the evil, evil. Or, as Aristotle had it, we can judge the character of a person by what his or her goals are. Judging by our only frame of reference, which is human, we can say that if God's goal in creating the world is to do it in such a way that his creatures can experience pleasure, which is a good, then we have another reason to consider God good.

(3)

Once again, in a Greek mode, if the gods are good, they did not create humankind to walk on two feet and to reason, only to press us to our knees and make us grovel, but to stand erect, reason, do good and deal with obstacles. Any storyteller knows that without obstacles for the characters to overcome there is no story, and I assume the Creator would be quite a storyteller.

(4)

We are Utilitarian in that we consider pleasure good, as long as the way it is enabled is harmless, and pain bad, other than as a warning sign of a physical problem. In our terms, pleasure is a natural reward for fulfillment – whether pleasure of the mind, body, or spirit, as a mental reward, physical reward, or an emotional reward. Pleasure is also a reason for the existence of life that posits a benevolent Creator.

(5)

Sacrifice with pain or at the risk of one's own life to help preserve another life is heroic, but the rational justification for the sacrifice ought to be to preserve the other life for the pleasure we derive from doing so and the pleasure the person we save will be able to experience; if for pain, the effort becomes less forthrightly understandable as good.

(6)

Since humankind is part of Creation, we may logically posit that we are meant to live in consonance with it. Since the workings of Creation evolve and nourish life but also terminate it, our natural adjustment is to be primarily life-directed and adjust to the concept of lifespans.

HAPPINESS

If life is meant to be good, we are meant for happiness more than for sorrow, so happiness results from enabling life as it has been gifted.

(2)

Creation is either indifferent, malevolent, or benevolent.

Since life has evolved and continues to be supported, we cannot consider Creation indifferent, at least, to the overall existence of life. Since life seems to have evolved each form of life with capacities suited to it, we intuit a positive intent.

If we decide life has evolved to be painful, we must conclude that Creation is malevolent.

If Creation has downgraded life to be a punishment though many generations – even if the fault, seemingly minor, is supposedly ours – we must think the less of Creation, because such an action against one's own creations does not express wisdom or goodness even as humans can understand and advocate those virtues. Since Creation is responsible for how we have been created, this inference is even clearer.

If we wish to understand Creation as benevolent, at least, in its life-evolving activity, we must conclude that life has evolved to be a blessing. This conclusion enables a praiseworthy vision of Creation. If we accept it, we ought to accept life with gratitude and vow to live in ways that are worthy of its blessings. Then life may attain its ultimate blessedness, as a gift of joy to each, a joy for all, and as our most credible praise to Creation.

(3)

Behold the resplendent ensigns of the joy nature takes in life – such as the gold sun, blue sky, emerald seas, green trees and colorful flowers.

(4)

The overall goal of Creation seems to be the fulfillment of the life-oriented potential of the universe and it seems to delight in the process, as apparent in, for example, a field of flowers. We should live in harmony with Creation's overall goal.

(5)

When momentous things happen, Epictetus recommends that we "remember how we feel when we hear that it has happened to others." But our primary focus ought to be on our own lives, in their heights and depths. Then we may, working from a solid foundation, become commendably involved in the welfare of others.

(6)

We are like musical instruments that life plays throughout our ranges, from the highest notes to the lowest; that is, we have evolved so that we can respond to the full range of experiences that existence presents to us. As a result, we may experience our entire potential for life and, if we are wise, we will, during the time we are fortunate enough to experience a share of the gift of life, sing life's complete song.

(7)

What is more important, Being or Becoming? Is it not, as usual, wisdom to balance them, so we may live peacefully and joyfully while we strive to achieve more.

(8)

Let us consider how today's culture militates against happiness. We wake up in the morning, energized and expectant. Then we learn from the news that, for example, 200 people were killed in an explosion in a distant land. We sense the implied question, "In the face of such news, how can you be happy?" Then we hear that a man in a city just killed his entire family. Again comes the implied question, "Don't tell me you're still happy?" Next, we find out that a renegade regime is attempting to acquire nuclear weapons, and we sense the more intense question, "What do you mean, you're still happy?" Then we're told that millions are starving in a famine-ravished land, and hear the even more intense question, "What do you mean, you're still happy?"

But then we chance to look outside. The sun is shining, people are going about their generally peaceable and goal-oriented lives, the other animals are going about their lives, and, most conspicuously, there are no explosions, at least, within our purview, and no one is prowling the street with a pointed gun. We realize that our personal lives are different from our media lives. And we decide not to let the first one overwhelm the second.

In fact, while making a place for the news is an ethical responsibility, to allow it to subordinate our own lives would be a disservice to them. Now, we vow to live our personal lives, while we provide only a prudent space for our media lives. We have demanded, out of respect for our own lives, a space in which we can be happy.

(9)

Despite being told that we have an undeniable obligation to participate in the world's agony, life was not gifted for us to be steadily unhappy. In fact, to be so would be an offense against life. Far better to let the media, whose providers know that without crises there is little interest, stoke the fires of outrage. Let them fan away, but let us stay away, so we may do our lives justice and save a goodly portion of them from the inconsiderately foisted tribulations of an outrageously death-bent world.

(10)

Polls do tell us that the happiest people are the citizens of nations that are least caught up in the tribulations of the wide world.

(11)

As soon as we wake up, we're tempted to pick up the rocks that represent the current burdens of our lives and pile them on our backs, until we're overburdened with them and tense. But when we catch ourselves before we begin, we can decide not to pick them up. Then we can start each day unencumbered and deal with the burdens we must as they confront us.

(12)

The death culture somberly demeans living well with the Romanesque canard, it's nothing but devotion to "wine, women, and song." The three are actually a delightful combination, and even traditional religion has managed to retain wine and song. But, of course, the attractive trio are only part of well-balanced fulfillment.

(13)

Attempting to be happy all the time is, we know, unrealistic and can, in fact, make us unhappy. We're bound to fall short. Let us be happy as often as we can and accept that periods of unhappiness are inevitable and can help us appreciate our happy times.

(14)

How hopeless our plight would be if, despite the liabilities of the death culture, we were all as happy as can be. We understand that our unhappiness grows out of sorrow over the depredations of the death culture, but it also grows out of our knowledge, conscious and unconscious, that the death culture deprives us, to a withering extent, of the lives we are born to enjoy.

(15)

We often hear that only simpletons aim at or can experience happiness. Yet the wisest, from Aristotle to Mill and Russell, have maintained that happiness is the goal. The question is, what leads to happiness? Aristotle's usual answers were, of course, that it results from life according to reason and by avoiding excess of any kind. We think much the same but, specifically, that happiness is most likely to result from life fulfilled in a wisely well-balanced way – in regard to the mind, through knowledge, reason, and imagination; for the body, lovemaking, fitness, other physical achievements, and dining; and in terms of the spirit, communion with the self, others, and all of Creation.

(16)

We can only be as happy as we might be when we consider the happiness of others, because true happiness arises from fulfillment, which is individual and mutual.

(17)

What is nobler than the quest for happiness? The quest for greatness, in individual and mutual achievement or in service to other people and other life. Yet even such activities grant us the happiness that comes from knowing we are pursuing life-directed goals.

(18)

There is, of course, necessity, but also nobility, in the well-proportioned quest for the emoluments that support our opportunity for fulfillment, namely, food, clothing, and shelter.

(19)

We include in the quest for happiness all of our most distinguished efforts at individual achievement and in helping others toward achievement. Even where great sacrifice is required, there is, hopefully, a modicum or plentitude of joy.

(20)

There is a direct relationship between the extent to which we esteem life and enjoy it and the extent to which we demean life and suffer it.

(21)

We strive to live more happily than we are encouraged to by contemporary society. We would obviously find the quest easier in a society that does more to enable happiness.

(22)

Many activities that make us happy or nourish our spirits so that we may continue our quest for happiness are readily available, for instance, the enjoyment of natural and humanly created beauty.

(23)

The more we appreciate the potential of life for mutually considerate happiness, the more we realize that we need nothing more than the realization of it.

(24)

Creation or, if you will, "the gods" have evidently done all they can or intend to do toward our fulfillment and happiness. The attainment has been left to us. Creation can observe how we're doing through our own self-awareness. While I hesitate to speak as if I can have any knowledge of "the gods," should they happen to look in, I assume they may distressed by our present behavior but know that we have the capacity to do better.

(25)

The world is widely sorrowful but improvable. Much of our grief grows out of liabilities we can alleviate, principally destructive behavior and underdevelopment. Excepting victimization by the first or debilitation by the second, we finally realize that the overall tenor of our lives, happy or sad, is usually the result of our own efforts.

(26)

Life and its natural joys are the most potent remedies we have for death and sorrow.

(27)

Let all who know the essential goodness of existence and the happiness inherent in it commit to the realization of our individual and mutual happiness as faithfully as we can.

(28)

In a modern society, where the necessaries of life are pretty generally provided, society creates a great deal of its own happiness or misery. Since society will only change when enough people in it do, we must strive to enable our own happiness.

(29)

There is a great deal of natural joy in simply appreciating the greatness of life and knowing we are each part of it.

(30)

The so-called wise may denigrate life, but the wisest have not overlooked our potential for happiness and some have observed that its existence must overall outweigh misery or the human race would have lost its enthusiasm to live and to bring new life into the world quite some time ago.

(31)

There is a great river flowing by called The Current of the Age, filled with the majority of people and events. The wise observe from the bank. We know if we wade in, we can be swept away by the overall flow, which today carries the multitude along on its generally death-determined and trivial way.

Firmly grounded, we can enable our own lives. When existence on the bank becomes more visibly inviting than it is on the river, we can assume that more people will chose to absent themselves from it deathward drift.

Those on the bank can, of course, offer help to those struggling shoreward. But we must always remember that our hope for fulfilling and happy lives requires us to remain out of main current enough so that we are not swept away.

We can, however, slip in from time to time to achieve a prudent goal. But it would be imprudent to wade in carelessly.

Many will attempt to help save others to the extent they can. Yet there will always be a proportion of heroes who will enter the river and do their best to help whomever they may make it to shore. They are those among us who are best in vivid action, just as there are those who are best in contemplation, planning, and overall effects.

We salute the life-devoted bravery of those who dare, the talents of those who cogitate, and appreciate the presence of those who help make up the generally happy populace.

(32)

The more rational and sensitive you are, the more an irrational and insensitive world is likely to upset you. The dynamic necessitates a degree of philosophical distance to ensure the happiness and the very survival of many who are the most capable of making a positive contribution.

(33)

The widespread effects of the death culture and consideration of what is being lost because we do not live in a life culture have afflicted much of my life with an undercurrent of sadness, but the realizations I have just shared have helped me lead, not a life lost or totally distracted, but one that has been, among many ups and downs, generally placid, fulfilling, and happy. I can say that I am glad I have had the opportunity to live. I have known great happiness and I wish you great happiness, too.

(34)

Let us each appreciate, fulfill, and enjoy life as the naturally miraculous gift its attributes proclaim it to be.

UNHAPPINESS

If there is such a thing as "the devil's work," and we do not mean to indicate that we believe in such a horrific red-tailed fantasy, it's making people think less of life and turning them away from it. An observant look at the sadly perplexed expressions on many people's faces indicates that the work is being done with incessant determination.

(2)

When we pass through rough water, it's wise to keep a lot of space between us and the rocks.

(3)

When the engine of the world roars and would speed us off to join its mad cacophony, we must put in our clutch, so we can remain in place. Our clutch consists of our commitment to life and our determination to preserve the quality of it.

(4)

There is no harder time to count life's blessings than at a tragic time. Sometimes we must give ourselves time to heal. Then we can return to an appreciation of them. By not to doing so we compound our sorrow with the loss of our Faith In Life and do an injustice to the abiding greatness of life.

(5)

Unhappiness has, of course, a natural place in the wholeness of experience, but we must strive never to let it continue so long that it overwhelms the overall happiness of our lives.

(6)

Happiness and unhappiness are the high and low notes of our lives. We can only do our best toward facilitating a preponderance of high notes in the composition that expresses and will finally represent our lives.

(7)

There is life-affirming heroism is sustaining our Faith In Life through a grave tragedy or a sequence of dire events – to try to understand the source of each catastrophe, so we may prevent repetitions as best we can, and to affirm, despite our loss, the overall worthiness of life.

MISERY AND DESPAIR

At times of abject despair, we need to turn to the wholeness of our lives, including the people who are most tenderly involved in our lives, and even to the life of every cell of our being. Do we have the right to absent ourselves from the wholeness, from other people, and to complete the lives of every cell that is at our mercy and functioning with wondrous brilliance to sustain our lives? I think not. We need to see our own life as an embodiment of the life of Creation and vow once again to be worthy of it as the triumphant gift it continues to be and enlivens us to continuing being.

(2)

To find ourselves with the irreplaceable privilege of living and to be in constant misery is a contradiction that, we hope, resolves, through growing awareness of the greatness of life, to at least a modicum of contentment.

(3)

Most preventable human misery grows out of irreverence for life.

(4)

The preventable forms of human misery have for some decades outnumbered the unpreventable ones and have usually been the more destructive.

(5)

Live for life and never let the death culture persuade you, through its overt or subtle ways, to vitiate or terminate your own.

REPRESSION

Repression often instigates expression.

LIVING IN PEACE

When we are at peace with life as a complete and sacred gift, we can be at peace with all the answerable and unanswerable questions of life.

(2)

The greatest challenge before the human race is learning to be at peace with mortality and in finitude, so we can embrace life as it has been gifted.

(3)

To facilitate living at peace with life, here are questions to which the answers seem to me self-evident. Do we need to know more than we can in finitude and is it practical to hope to? Can't we know enough through the natural appreciation of life, which includes a healthy adjustment of our place in the unfolding of the universe? Do we really need, and is it logical to want, more than life itself can offer us? Do we need or actually want to live forever? Does anything in the universe last forever? Do we really need and is it appropriate to expect a personal communication with whatever made it? Can there be a more direct communion with what made it than through life and our life-enhancing participation in it? Don't we have before us what has been made, and aren't we all part of it – in fact, part of the embodiment of its animate life, which is its highest achievement? Don't we detect in our own lives Creation's devotion to life and can't we find a satisfying goal in living in harmony with its devotion. Are not logical answers to these questions sufficient to salve our souls and make us feel privileged to have a share of the life of this life-blessed orb in the ever-active and majestic universe?

(4)

The best way to make peace with the past is to appreciate the present. The crest of the wave is only where it is because of every molecule, or event, in it.

(5)

We can make a placid place for being by reaching out with our imaginary scissors and cutting the cord that pulls us forward toward becoming so intensely we cannot stop to enjoy our lives. We can also reach behind us and cut the cord by which we have our baggage in tow, so that we're less encumbered by the past and freer to enjoy the present. We can enable ourselves, freely but responsibly, to be our best.

(6)

We can make ourselves anxious when we subordinate the wholeness of our lives to any single event or goal, and doing so is a disservice to our whole lives. When we appreciate our lives, we're more able to retain a steady vision of the wholeness and keep each part in perspective. Such a peaceful perspective also helps us see more clearly that living for the moment is a prescription for making ourselves anxious. In the moment may lie all possibility, but our lives are made of innumerable moments, which the wise don't calculate with anxiety.

LOVE RELATIONSHIPS

A good love relationship provides, within mutual feelings of love, what may be described as a nourishing coziness, instead of the disheartening loneliness we can experience when we're on our own, and such a relationship can nourish a happier, healthier, and more productive life.

(2)

A love relationship requires support from both sides, like a bridge.

(3)

Two values ever apply to love relationships: individual and mutual. We each balance them as well as we can.

(4)

I sometimes say I only know two things about new relationships for certain, and, specific relationships, being cantankerously variable, both are general perceptions. If a new relationship seems to come with lots of answers, instead of lots of questions, it may have exciting possibilities; and, if the person is there, instead of leaving us to wonder where he or she may be, love may be in the offing, because love manifests itself as attraction.

(5)

All of our experiences, including the love we can express, grow out of us as individuals – a perception that alerts us to the importance of leading fulfilling lives ourselves, so that we may genuinely enable fulfilling lives for others.

(6)

Self-love enables the true expression of love to others; what can we truly express that we do not have within ourselves?

(7)

Opposites may attract visually but, temperamentally, there can be problems; for instance, a voluble character may perturb a tranquil one, as a waterfall still water.

(8)

Having a deep relationship eliminates or limits having superficial relationships, while having superficial relationships eliminates or limits having a deep relationship. We each choose as wisely as we can.

(9)

While variety can be enjoyable and satisfying, especially when we're young and exploring or when we simply haven't found someone we deeply care about, ceaselessly wanting more relationships, even when we're fortunate enough to be in a good one, obviously endangers a good relationship, but also indicates some devaluation of individual value, while settling into one relationship can help preserve it and indicate individual value but, when a relationship is unhappy, simply lethargy.

(10)

When one of the people in a relationship longs to be free, a period of it may help him or her better appreciate the relationship.

(11)

It is amusing to observe the volubility of people who just meet and the frequent silence of people who have been together. Obviously, the first are just learning about each other and they are eager to hear what they each think about many topics, while the second already know a great deal about each other and generally only talk about topics that are new to them. The initial talkativeness of people who just meet partly allows them to invest each other with themselves, to see if each might be a ready home for the other's persona. If the answer appears to be yes, they value each other because they sense that they exist in and are valued by each other.

(12)

A good love relationship is a foundational aspect of each person's fulfillment and a treasure, whether its lifespan is coincident with the length of our adult lives or part of their length.

LOVEMAKING

Lovemaking, when accepted and enjoyed as a natural part of life, contributes, of course, to our entire well-being. It is a joyful mutuality that, when excessively repressed, amounts to what we may call a partial death of our life energy. As a repercussion, our will to live may afflict us with negative feelings and fantasies of suicide.

(2)

Being comfortable with and appreciating lovemaking is a fundamental aspect of being well adjusted to life.

(3)

The voice of Creation is ever within us, and it's most urgent need is that we procreate, so that amid the ups and downs of our development, we may at least continue. The voice is not to be repressed; it breaks through and, when not allowed to do so in natural ways, may escape in perverse ones.

(4)

The death culture loves to flirt with lovemaking and then punish those who succumb in any other way except the narrow path in which it is excused. We would never go so far to say the death culture ever does actually approve of it, let alone understand that, since it is the way we create new lives, it is the fundamental physical sacrament of life.

(5)

I'm disinclined to excess in expression, because it intimates illogic, but how lovemaking, either for joy or joy-plus-procreation, ever came to be considered anything but sacred seems to me a diminution of it by the death culture that can only be called sacrilegious.

(6)

Lovemaking still struggles under the burden the morality of the death culture heaps on it. The cost can be seen in sadly denied lives and the relatively meager consolations of pornography.

(7)

The death culture is so maladjusted to life it has reduced the beauty and appeal of physical life to such unworthy displays as peep shows – where its deprived victims actually pay to peep in on natural human physical life as it has evolved and was originally conceived by the Universe itself or by a Creator or Creators beyond our experience.

(8)

The oft-expressed anxiety that people in a naturally free society would make love to the virtual exclusion of other productive activities, most especially what we call work, reveals a shameful mistrust of Creation's design intelligence. It is also based on the illusion that we have an infinite supply of physical energy. Most people finally do experience satisfaction and, if they go on, exhaustion, and finally the longing to do something else. Creation has endowed us all with a naturally well-proportioned distributor of our life energy. When, however, an aspect of it is repressed, it can loom larger than it is. Imagine, for example, if society's inclination was that we should repress thinking. Then would not the great hidden desire be to think?

(9)

When we treat something as clean, we help make it clean; as dirty, dirty. The process of making things in our image of them applies to many things.

(10)

I often refer to the language that we've inherited to describe our bodies and lovemaking as a legacy of scorn. Among the demeaning and demeaned terms we find "come" as the moniker for seminal fluid. It is astonishing that we must still right our views of even the most obvious graces of life. Seminal fluid is Creation's medium of procreation and of our continuance. As such, it is sacred to life and should be held in awe. Then we will realize its presence never reduces a man or a woman but, sticky as it may be, enhances them and, when washed off, leaves no residue of shame. Thoughtful lovers, even if they go their separate ways are, if anything, made cleaner by this half-medium of new life. And, speaking of the media of new life, let us also affirm that the menstrual cycle is a sacred aspect – ova, endometrium, and blood.

(11)

Sexual repression, excessive possessiveness, and trembling anxiety over infidelity often go together. We could increase our mutual well-being substantially by, as the saying goes, lightening up a bit.

(12)

The body knows more in its inclinations than we do about it. What the mind best adds is prudence and consideration for others.

(13)

There are two basic ways to make life thrilling, especially physically – fulfill it or kill it. So far, we have excelled far less at fulfilling it. Part of the reason is that the death culture limits fulfillment, particularly physical. But making love is, when consensual, perfectly moral, while making mayhem is, when not defensive, perfectly immoral.

(14)

The continual stress we experience by living below our natural lovemaking inclinations, a condition we refer to as frustration, takes an incalculable lifelong toll on our health and very likely shortens our lives. In the very young but physically mature the entire denial of it may account, not only for some pimples, but for self-destructive behavior. Of course, excess also has its liabilities. As usual, moderation, with perhaps occasional joyful excess, seems not only logical and wise but naturally inviting.

(15)

Beliefs that are in harmony with naturalness allow us to partake of life as it has been gifted, or in divinity as we can truly know it – where balance is everywhere evident.

CONTRACEPTION

Facilitating population moderation should not be difficult but we make it so in many ways.

There are the obvious constraints administered against a commonsense approach to contraception, including outright prohibitions. Such policies can only be enabled through insensitivity to the depredations attendant to overpopulation.

We ought, of course, to endorse the use of condoms, birth control pills, and other means of contraception, but we ought to go further and foster the positive view that it is natural and praiseworthy to make love, not only for procreation, but also for mutual sexual enjoyment.

For instance, a commendable but evidently daring recommendation toward the moderation of our procreative proclivity might be, "Have Sex, Not Kids."

ABORTION

These days, a commitment to the sanctity of life can readily be confounded with the Pro-Life movement against abortion. The possible confusion is part of the difficult situation in which we find life caught up. But being Pro-Life has a far wider significance.

(2)

Life itself can only exist as it was gifted when each person has the freedom to make his or her own choices. So, despite the regrettable anguish of the situation, I believe an aspect of freedom includes the right to choose an abortion. It seems to me that a mature life has the right to decide what new life she wishes to give birth to. A fetus is not yet an independent life; it requires a woman to support it. But the woman is an independent life and has the right to decide how she will conduct the life of her body.

Certainly, education and contraception are to be preferred, and they can coexist with the right to choose, but a safe abortion cannot coexist without the freedom to choose it.

Finally, we must choose between a fetus not coming to term or a woman and the fetus perishing at the hands of an unqualified abortionist. We choose to save the woman from such a fate.

STEM CELL RESEARCH

Life-enhancing science has an inviolable charter. It grows out of our potential to work with and improve life as it has been gifted. So stem cell research should be, not only allowed, but encouraged.

Ethical objections generally arise from considering an embryo in the abstract, as a living person in miniature. This view – typically of men, but also at times of women who may not appreciate their unique contribution to the development of life – neglects the role of the female in providing a nurturing uterus for an embryo to develop, so that, in time, it may become an independently alive human being.

Without the ministrations of the woman's physiology, the embryo has no chance of developing into an independent life. It is destined to the bin behind the sign we often see in today's healthcare facilities: "Biowaste."

No one is, of course, suggesting that embryos be removed from pregnant women for the purpose of stem cell research.

Stem cell research on embryos that would otherwise be destroyed enables them to be beneficial to life, just as postpartum humans can, at their best, be beneficial to life. So we ought to be, I think, guided by the belief that life at any age is better employed in the service of life, not wasted.

Of course, we would all prefer ways to do all the stem cell research we feel would lead to improved healthcare without the necessity of destroying embryos at all. The possibility appears doable.

THE HUMAN BODY

The processes of the body create and support our lives – including our consciousness, thoughts, feelings, and actions. When appreciated for all it is, the body merits our devoted reverence and care. It is the very embodiment among us of the design and wisdom of Creation.

(2)

Your body contains your life and its processes produce your life. Let us do it justice and recognize that it is sacred to life, in fact, it can be seen as the very tabernacle in which each life dwells.

(3)

Understanding purifies every natural aspect of life; appreciation sanctifies it.

(4)

The body is the form, substance, and host of life as it has been created. We might well see it as the tabernacle in which Creation has placed life.

(5)

The body, mind, and soul are one – the trinity of our natural animate being. While many disagree, we have numerous indications that the unity is true. For instance, as J. S. Mill and others have noted, when we become ill or older, our bodily decline affects the quality of our mental lives. We note that it also obviously diminishes the natural liveliness of our spirits.

(6)

Life is of your body, its ongoing product. So we owe our daily lives to our bodies. When they are hurt, our lives are and, when fatally, our lives are over.

(7)

It helps us to respect our bodies to appreciate that, at least in this life, our bodies and our being are one. Our bodies are who we are entirely. This perception can help us overcome the tradition of disrespecting our bodies as mere earthly matter, while our eternal souls are merely residing there for the short term.

(8)

We might consider our bodies as the lifelong creator of our daily lives, ministering to our metabolic needs and responding to our volitions with ceaseless devotion.

(9)

Besides supporting our lives as we are conscious of them, the body conducts innumerable processes we are not aware of but depend on. The wisdom of its workings is equal to the requirements of life and many of its processes are obviously still beyond our understanding. The more we do understand its workings, the better we'll be able to appreciate and compliment them, particularly in terms of healthcare.

(10)

When we realize the primacy of Creation's intent that we survive, we understand why the human body is the most attractive existent we can see, touch, and desire.

(11)

Sensation itself is such an astonishing achievement we ought to consider it a natural miracle, which we're privileged to experience. Aptly fathomed, sensation is so invaluable we can appropriately be ecstatically adulatory over the tender touch of another human being and express some appreciation even for an itch.

While we will certainly enjoy the loving touch, we can, of course, still be annoyed by the itch and scratch it.

(12)

Your body is the foundation of your life; the better you care for it, the better you can build on it.

(13)

Since the body is the handiwork of Creation and all of life depends on it, love it as foundational and then its products of mind and spirit.

(14)

When we regard the human body with respect, we enhance it; with disrespect, we diminish it. The same applies to lovemaking and, in fact, to anything else.

(15)

It is logical to live in awe of our bodies and to love them.

(16)

How our ancestors, to disown the body so, as Plotinus did, must have suffered physically, when, rightly appreciated, it is transcendently capable and astonishingly beautiful.

(17)

The human body has evolved with such balance, beauty, and capacity that the ultimate source of it obviously created it with surpassing intelligence and diligence and intended it for distinguished achievement and appreciative enjoyment.

(18)

Let us reflect on the body as we ought to: My body is the temple of my life, and, within the span of my life, I may build with wise fulfillments an altar within it, as a symbol of my praise for the gift of life, my contribution to its distinction, and my worship of its Creator, natural and, if an existent, of its extra-natural Creator or Creators.

(19)

Your body, with all of its attributes and capacities, is actually your greatest personal treasure. Without it, how could you be in life? So do take intelligent care of it.

CLOTHING AND NAKEDNESS

We cannot be half for life but for it in all of its aspects. Ours is not to qualify it but to appreciate it and conduct it wisely. And, while clothes are a wonderful comfort, protection, and delight, we do have to remind ourselves from time to time that we were not born with them and are the only creatures who find them necessary.

(2)

We owe our lives to the parts of our bodies we're conditioned to be most ashamed of. They are the natural source of each human life and so of our past, present, and our continuity, along with our workaday and intimate feelings of unity and joy.

(3)

We know, as we do so, that in removing our cloths, with volition, we are, to an extent, slipping free of our inherited shame.

(4)

People who choose might go without clothes more often than many of us would dare. Imagine, for instance, how some hardy souls have conditioned themselves to swim even in winter's icy waters. People who prefer to live as naked as we were created might wish to establish towns and cities, and the only objections others might have would arise from shame at the way we were created. I must confess, however, that I myself, as the result of conditioning and my lifelong perception of comfort, would rather don my clothes at the usual times and live in a society that shares my preference.

(5)

It is perfectly natural to desire to see and touch the greatest physical gifts Creation has given us, as we can experience them in ourselves and with other people. Of course, along with the acceptance of the gifts of life and the natural rewards attendant to them, we must acknowledge that life can only exist as gifted when individual freedom of choice does.

(6)

Naked or clothed, we have harmed and killed one another. The change we need is not so much about clothing or nakedness but far more about the appreciation of every human being and, in fact, every creature as a naturally miraculous embodiment of Creation's own life.

PRIDE

When we appreciate the grandeur of Creation and the fulfillment that is within our capacities, we can be proud of our place in Creation and, when our conduct is worthy of our life-oriented potential, proud in a healthy and satisfying way.

DIVERSITY

Nature so loves diversity that no two creatures or things are entirely alike.

INDIVIDUALIY

Life itself resides in the individual. So intelligence, action, and love reside there. Our goal is to enable the individual, who we take as, not only a singular, but a social being. When we express the finest inclinations of the mutually considerate individual, we enable the finest potential of life, individually and socially.

(2)

While society helps mold the individual, the individual can help change society.

(3)

The world in general has little to offer; one person, may much.

(4)

It is proverbial that the first way to make a world we can be proud of is to make ourselves a person we can be proud of.

(5)

Our logical first interest is the preservation and fulfillment of the self. The belief is fundamental and practical; for instance, when we fly we are told that, if the cabin loses pressure, we should put on our own oxygen masks first, so we can we stay conscious to help others. Our second interest is the preservation and fulfillment of those we love. Our third for friends and then acquaintances. Until, by coronas of care, we encompass all of life.

(6)

The individual finds most commodious the government and economy that enables his individuality. Once he has them, he can enable his fulfillment and, to the extent he chooses, the fulfillment of others.

(7)

I know of no greater prospect than the individual in the service of life and then a society of such individuals.

THE EGO

The ego can be seen as the overall way a life says he or she exists, and our behavior as made up of the many ways we say it.

MUTUALITY

We survive, flourish, and enjoy life only with a wise and easygoing balance of individuality and mutuality.

(2)

Behavior in agreement with life cannot be selfish, because life cannot achieve its greatest potential with self-centeredness as a guide.

(3)

Human value depends, to a great extent, on fellow feeling. Without it, human value seems something to scorn; in fact, we cease to be worthy of the liabilities we place on the other creatures of the earth for our continuance.

(4)

The purpose of society should be to enable the individual and society as a whole to live fulfilling lives.

(5)

People reveal themselves readily through what they say about others who are not present and how they treat others who are, especially those who are under their influence.

(6)

Consideration usually nurtures consideration and thoughtlessness, thoughtlessness.

TRUST

When we have more trust in life and the wisdom inherent in it, we will trust ourselves and our wisdom to live it rightly.

KINDNESS

Kindness may be the most commendable virtue, even more than love. It invites behavior that is only commendable.

(2)

To be kind has wonderful advantages but is not without disadvantages; for instance, as an expression of it, we allow people to become who they are and may be hurt by them because they abuse the freedom we have granted them. Yet we like being who we are and, rather than being unkind, prefer to hope for others who are kind.

(3)

Kindness alone could change the world.

HATE

Hate is only possible when we little value the object of it. When we value life, we feel awe and love for every embodiment of it and can only bring ourselves to dislike whatever diminishes or harms life.

GUILT

Life libeled with infused guilt encourages dismissal of it and dependence on an unknowable God for redemption. Life graced with healthy pride encourages acceptance of it with gratitude toward the universe and any God or Gods outside of our experience of it.

FORGIVENESS

Forgiveness is a two-way balm – a welcome relief to those we forgive and to ourselves, and a compliment is the readiest way to mollify an antipathy. Failing to win over an adversary with both, our best choice is retreat; our final option, defense.

GOVERNMENT

Since intelligence resides in the individual, the individual is the only resource capable of realizing the potential of Creation. So the best government is one that enables the individuals who make up the society to achieve their utmost. When a government subordinates them, it represses the best promise of its own society.

(2)

The best government is one that protects and enables its citizens to fulfill their lives, while it also helps other nations to achieve the same goals.

(3)

When most of us think of the vote, we turn to election times. The everyday vote we each have is how we choose to spend our time and money.

(4)

The greatest danger to a happy society is unreason in power.

(5)

Is the degree of intelligence and knowledge with which America and other nations conduct their everyday and international affairs equal to remedying our problems and, further, to establishing our answers?

(6)

While a strong central government tends toward the aggrandizement of insiders and the subordination of individuals, there is hope that, in a society guided by the care of life and its possibilities, a strong central government would retain the awareness to work toward the enablement of its individual citizens. The only possible realization of a society's best potential is can be summarized as a policy of individuals first, society second. Yet, even given such an enlightened society, the populace would necessarily retain the fundamental safeguard of the right to vote for its leaders.

NATIONS

The greatest overall achievement any nation could give the rest of the world is a demonstration of the benefits of life-affirming conduct, enabled by sensitivity to the greatness of life, with the concomitant imperative to help preserve peace.

(2)

A nation is exemplary to the degree to which it is devoted to the care and fulfillment of life.

(3)

While nations do protect diversity and subdivide responsibility, is the air divided into nations?

AMERICA

The genius of the America is not military, except in defense. The ruthlessness required for other types of military activity is not consonant with our ideals of freedom and democracy. Our genius is economic. Since most of the world wants economic success, we can exert our best influence by helping to enable it for a world of nations. Since freedom is more attractive than governmental dominance, the ideology of America provides its economic assistance with an inherent advantage over governments that inhibit freedom.

(2)

The treasure created by American workers is wasted on misdirected overseas adventures, with the intent of exerting American influence, when, invested at home, the same treasure would help create a nation that is far more influential because of the emulative quality of its life. Yet our military exploits leave only meager resources for us to become an exemplary nation.

(3)

Americans want to do justice to their highest hopes but need leaders who understand what those hopes are. Yet the public does have a duty to remain enlightened and recently what many Americans want is regressive. A widely ignorant populace cannot lead the world in governance, economics, or in any other meritorious manner.

(4)

America's Founding Fathers, who were generally inspiring and informed devotees of the Enlightenment, might today be described as libertarians with a social conscience.

(5)

History tells us that some of the most exceptional of our Founding Fathers, including Washington, Jefferson, and Franklin, were Deists or skeptics. This is a fundamental reason that the Constitution call for separation of church and state.

(6)

America has sacrificed enough for other nations and needs a period of healing and renewal.

A FLAG OF LIFE

I saw an American flag with a peace symbol where the stars usually go. I wonder what one would look like with a symbol of life placed there?

ECONOMICS

The economic system of a life-oriented society would develop out of the widespread appreciation of the value and potential of life and the consequent need to support the care and fulfillment of it.

(2)

A life-oriented economy would work to enable the individual, local and national fulfillment of its citizens, while it helps to protect their lives and environment.

(3)

Business leaders with vision seek to deliver a fair amount of shareholder value while they enable good lives for their employees and provide for more corporate success over time. Those without vision seek to over-deliver shareholder value while they undervalue their employees, vitiate their lives, and deliver less long-term success. Leaders with vision are always in danger of being replaced, because investors usually seek maximum immediate profit. So a company that performs well in the present and will do so in the future requires, not only a leader with vision, but a board of directors and key employees who have it.

(4)

We are often, amid the potential for plenty, exhausted by the quest for enough. The dynamic is one of the less desirable effects of competitive economics. Yet private property is the foundation of personal independence and therefore of independent thought and fulfillment. When a government owns all of the property, the majority of the citizens are reduced to being life-denying knaves and slaves.

The more we know, however, to live for life and its natural rewards, with the knowledge that life flourishes first in the individual, the more a cooperative economy that honors individual freedom may become possible. Since all previous cooperative economies have been organized for material production or different utopian ideals, we have no experience with which to make a credible estimate of its likelihood for success.

(5)

The misallocation of a nation's resources cheats the citizens of the lives their talents and labors merit. It is a violation of their welfare, potential and happiness.

(6)

All the chains are gold: the more we want, the more we enslave ourselves to pay for it. Yet we need material well-being for our fulfillment. So we ought to settle on a prudent middle ground that can enable a pleasant life while we're free enough to fulfill as much of our unique potential as we can manage.

(7)

Intelligent life support allows for a naturally rewarding life more than diligent acquisitiveness. The necessities of life and a modicum of creature comforts help enable fulfillment, but an excess of material wealth cannot be a satisfying substitute for it. We need only observe the desperate doings of wealthy individuals who are not focused on genuinely rewarding fulfillment.

(8)

Today, the wealthy are able to go far more beyond survival and devote themselves to the independent fulfillment and enjoyment of life. But they must do so in a world that largely does not understand their goals and in relative isolation. They also may not be wise enough to choose life-enhancing goals. So their satisfactions are seldom what the human spirit longs for. Many of them would find more satisfaction as members of a life-oriented society with appropriate life-enhancing values in place and embraced.

(9)

Nature appears to distribute common needs widely, such as sunlight, air and water, at the same time it distributes them opportunistically; for example, it distributes our atmosphere generally, while it only allows such a life-sustaining atmosphere to occur on planets that are suited to it. It also distributes specific items opportunistically, such as growing types of plants where they most appropriately flourish. The latter enables our ability to distribute specific items more generally.

(10)

Many would find more fulfillment in a workweek made up of three days, with four days for independent life, so that the proportion of days is toward independence. Yet many others, who find great fulfillment in their work, would obviously still prefer to spend more days at it.

(11)

Although an effective government must be enabled to control its finances, it would be empowering if we were able to indicate on our tax returns how we prefer a portion of our payment allocated, for example, to education.

(12)

The fundamental problem with cooperative economies is that they subordinate the individual. One has yet to be devised that would enable the individual. Such an economy would support naturally fulfilling, mutually respectful lives and provide for common needs, like roadways, defense, and protection of the environment.

But we do not yet have a sufficient grasp of the rewards inherent in abundant post-material incentives to compel us toward such an economy. The description and implementation of any such economic system must wait until enough people understand its potential to establish a significant and fulfilling society based on it.

The quality of such an economy, like every endeavor, will depend, not only on the expertise we can bring to it, but on the character.

Given the evidence to date – that cooperative economies invariably repress individual freedom – most wise and wary souls prefer the independence inherent in competitive economics, which does, to its often-unappreciated credit, enable the individual.

Whether we approve or disapprove of competitive economics, today, as in the past, the only workable foundation for individual freedom is a free economy.

Many quite intelligent thinkers seem to forget that the first to be silenced by political or economic tyranny are the most independent-minded.

(13)

The economics of scarcity can, through global development, technology, and life-directed priorities, give way to the production of plenty.

THE SYSTEM

The goal of a life-oriented system would be to provide enough material wealth, technology, work, art and entertainment for every person to lead a satisfying life.

(2)

Since we do not understand life entirely, we cannot create systems that nourish it completely. We can only give priority to systems that have the most ability to enable, in an advanced society, the naturally fulfilling and mutually considerate inclinations of life.

(3)

Tyranny and poverty result from debilitated systems of government and economics. Whether we like the concept of a system or not, we who are fortunate to work in functioning systems ought to appreciate the extent to which they enable our possibilities, not only to fulfill our own lives but to help other people create the conditions that will allow them to fulfill theirs.

(4)

We all function within a universe of systems, not simply one composed of political, economic, and military components. For instance, when we drive, we are operating within the systems of the car, other cars and vehicles, roadways, traffic lights and laws, air, gravity, and, actually, the entire system of the universe, which helps support the stability of the earth.

(5)

The more systems sustain us, enable our best intentions, and protect us, the better we are likely to consider them.

GLOBALIZATION

Since prosperity is universally recognized as superior to poverty, globalization is inevitable. The challenge is to effect it in as broadly a life-enhancing manner as possible.

POVERTY

Life as a self-satisfying fulfillment obviously depends on sufficient material well-being and education. The more poverty and ignorance are overcome, the more numerous will be the people who can appreciate the value of life, want to care for it, and find in it their fulfillment.

(2)

Many people who are called homeless are incapable of leading independent lives and can be more accurately described as "institutionless." They need a more caring society.

SELF-INTEREST

Self-interest in excess contradicts its goals, which it only achieves in a balance of self-interest and mutual interest, because both comprise our fulfillment and are reciprocally nourishing.

A SOCIAL CONSCIENCE

A life-oriented society would provide amply for the lives of its weaker members and help alleviate liabilities against life worldwide but would also have the wisdom not subordinate the lives of its most productive members. Light loads help the fastest horses win races.

VIOLENCE

The effect of violence on the body is proof that violence is wrong, except to protect or save life.

(2)

Under excessive life-denying conditions, including stifling repression or murderous poverty, flare-ups against the offending force are inevitable. If they were not, one would be more concerned about the vitality of the human spirit and its innate will to live. We can only hope the remonstrations are nonviolent.

(3)

Life-destroying explosions are rattles of the death culture that remind us how death-directed and insensitive to the greatness and sanctity of life many still are.

WAR AND PEACE

War is our most catastrophic expression of irreverence for life.

(2)

War is an orgasm of the death culture; peace, of the life-culture. War is about all-out killing; peace should be about all-out living. Then we can best hope to attract people to life and repel them from war. Equals in excitement most effectively counterbalance each other.

(3)

Nation against nation, class against class, and religion against religion are calamitous for our fulfillment and happiness. Diversity within unity provides for these values and invites us to participate as joyful individuals, spiritually united.

(4)

The prospect of a nuclear war, while terrifying, is partly sustained by the erroneous belief that all of the weapons will work. But probability advises us that anyone launching a large number of missiles will likely find a certain proportion of them striking unintended targets and some, in fact, falling back down on the places from which they were launched.

(5)

As to who starts a war, a volcano may blame the jettisoned rocks, but it is still very conspicuously the cause of the eruption.

(6)

Peace, as long as it preserves life, is always an honor, especially when initiated.

(7)

Humanity, through insensitivity to life, destroys it – and, most catastrophically, in war.

(8)

Peace can be won but only by our appreciation of what it kills and our determination to prohibit it.

(9)

War, in a life-nourishing society, would be contradictory or defensive.

(10)

War cannot coexist with the worldwide care of life.

(11)

Let us hope that one day war will be only the gore of yore and that peace will prevail forevermore.

HEROES

We have always been graced with heroes of life, as in those who save other life at the risk of their own. In a life-oriented society, we would have as many, likely more, and they would know clearly the inspiration for and merit of their efforts and sacrifices.

(2)

What are traditional heroes, rightly appreciated, but heroes of life, as, for example, the soldier who falls on a grenade, deciding to sacrifice his own life to save other lives.

(3)

Does a true appreciation of the greatness of life and our commitment to it mean that we will never sacrifice our lives for others or in a great cause? On the contrary, the reasons we value life inspire us to life-oriented causes. They prompt us to make heroic decisions but, rather than foolishly, by considering the values at risk and then acting so that we may help the most life, judged by its quality and quantity, live.

(4)

The largely unrecognized heroes of today are those who maintain intelligent, civilized, and productive lives in a world that widely demeans many life-enhancing ideals and dwells on ones that trivialize and destroy life.

(5)

While many suffer and pray for a better future life, a few alleviate suffering and enable a better present life. Who are the heroes?

(6)

When something good happens, clearly by the exceptional efforts of an individual or a group of people, many thank their God for the deliverance, but the more insightful know to thank the hero or heroes who effected the good event and know that in that priority they most credibly express their reverence to the Ultimate Source of the individual or group whose capabilities enabled the achievement.

HAIR

Our Founding Fathers and many who fought for the independence of America had long hair. So did many of history's greatest scientists, philosophers, and artists. But then there are also some, primarily more recently, with short hair.

RACE

Should The White House be renamed The President's House?

CRIME AND PUNISHMENT

When we value our lives more, there will be fewer crimes, not only because we will be reluctant to transgress on the lives of others but also because we will be reluctant to diminish our own lives. How repugnant, for instance, would the prospect of prison be when recognized as isolation from the possibility of fulfillment? It would signify and magnify the condemnation to death in life.

MEDIA

While media do us the commendable service of providing news and entertainment, the media in a life-affirming culture would have almost entirely different content than they offer up in our devotedly death-dominated culture.

(2)

The news intimates that it represents the majority of events in the world, but it only covers the most spectacular ones, as, I suppose, it must. It is up to us to look around us, realize how much more is going on than is in the news, and keep the spectacular events in their proportionate place.

(3)

Our personal lives are made up of events that begin within us and proceed outward in coronas of appropriate concern: first, to events between us and those we love; next, events between us and our acquaintances; then events in our own communities, and finally events, primarily presented by media, in the wider world. When we turn our attention to our own lives and the lives of those adjacent to our own, we give ourselves a ready antidote for the domination of them by the media.

(4)

Since social values are learned, the substitution of media for personal interaction can result in interpersonal maladjustment, a liability seen most conspicuously in students whose absorbing involvement in media inhibits their interpersonal activities and the development of social skills. Since personal interaction is the only way to nourish fellow feeling, they find themselves without it and, consequently, find no inner objection to striking out at their fellow students.

(5)

When we observe how the media build up public figures to involve us in their lives and then tear them down, achieving a two-fold interest in them, we realize even more the wisdom of Descartes' saying, "He who has lived well has hidden well." The smiling marionettes are trotted out and allowed to cavort and then, if they make one misstep, they are destined to become the troubled objects of plenteous scorn and are nailed to the cross of their wrongdoing – more often than not, a preposterous or relative peccadillo – where they must suffer before an incensed public.

Metaphorically, the procedure can be seen as follows: First, let us build the house; then let us tear it down; we know both will attract a crowd.

No wonder wisdom encourages wariness.

(6)

What kind of headlines would indicate that we're turning more toward life? Here are a few examples: "Government funds massive program to improve the quality of life in America." "Education, the environment, and elder care top Congressional spending bills." "Pentagon allocates majority of budget toward public works in developing nations." "U. S. International Corporations Put a Premium on Enabling a Better Quality of Life for its Employees." "World Celebrates Life Day."

(7)

Nothing stirs the worldwide crisis pot like the incessant churn of the overheated news media. Yet it is better to know the condition of the world, else how would we understand the magnitude of the crisis and effect means we hope will result in change? Our is to modulate the impress of media on our personal lives.

(8)

The increasingly widespread availability and sharing of information enables the hope of a rapid global Enlightenment.

ENTERTAINMENT

What is life-oriented entertainment? We have had some and generally know what it might be, such as works that highlight intellectual achievement, nourishing love, or mutually sensitive interaction. But it remains a relatively blank canvas, inviting our life-devoted inspirations.

(2)

Entertainment is best as a compliment to life, rather than a vacuous substitute for or, worse, an explosive diminution of it. I do not mean unremitting praise of life but an authentic depiction of the entire spectrum of life as it has been gifted, not only in its blessings, gains, and joys but also in its challenges, losses, and sorrows – the complete orchestration of life's sacred symphony.

ART

I prefer art that expresses Faith In Life, rather than infatuation with death.

(2)

Art is intelligence, rendered as verbal, visual, and musical splendor. While some of the energy that goes into its expression may be, as Freud maintained, sublimated sexual energy, art is a natural part of life and, logically, would be provided with its own share of our life energy.

(3)

Artists starving in a hovel are emblematic of the death culture. In a life culture, artists would vow to make theirs lives the benefactors of their talents, not the victims of them. It is a duty we owe to the greatness of life. And by doing so, we preserve our characters, instead of allowing them to be debilitated, so we are not embittered but retain our generosity of spirit. We may also achieve more, because we preserve ourselves from the stress of economic urgency, and so we have creative energy that might otherwise be lost to anxiety.

Finally, we allow ourselves to have healthier, and probably longer, creative lives.

(4)

What is the role of the artist in a mass media culture? He need not be overwhelmed with information or, in Marcuse's terms, merely reflect it. If he is sensitive, his contribution lies beneath the inundation of information. He observes the text of life but senses the unvoiced subtext, the voice we hear within ourselves and believe the content is only ours. He renders it, so we can all share it, in the relief of catharsis, and rediscover that we all have a great deal in common. Adapting Edmund Wilson's term, we experience "The Shock of Recognition."

WRITING

Writing, like all endeavors, is the fulfillment of a potential of life and is therefore sacred to ourselves and to all of life. So writing as well as we can is part of our religion.

(2)

To write as best we can, we must not allow ourselves to be victims of the bandied obstacle to it, "writer's block." Ironically, those who talk about the difficulty of writing often write the most. They are also being inexact. Writing is an effect, not a cause. First, comes thinking; then "writing." When our thinking is clear, our writing is merely the secretarial recording of it. If I can think clearly, "I want to go to the store and buy an apple," it is simplicity itself to record it. We can get into trouble when we think about writing as an end, because it is an effect of our thinking, and we cannot control effects, only causes. Sometimes, we can become anxious about our ability to write, which can prompt us to let writing get ahead of our thinking, and the effect of unclear thinking is, of course, garbled writing.

The way to easy writing is to be confident and patient with one's thinking; then, when it is clear, merely to record it. Writing, or more likely today, typing into a computer, is only the hands and fingers taking dictation from the mind. We need only be the good secretary of our thinking.

(3)

I write scientifically credible spiritual beliefs to help initiate a life-saving age.

MUSIC

Since music records the movement of the human spirit in tone – its rise and fall, smile and sigh, without ideational specificity, we already have a great deal of instrumental music that is life-enhancing. Mozart remains a wonder. We have some specifically vocal music that is life-oriented, such as Schiller's *Ode to Joy* as set by Beethoven. We hope to hear a great deal more of it, composed as the light of life dawns more widely.

(2)

We resonate with music because its highs, lows, and other inflections are consonant with and express the highs, lows and other inflections of our lives. Beethoven, sensing the effect, called himself "a tone poet."

HEALTHCARE

The inhibition of medical care by governmental or insurance agencies is unethical; the workable remedy would be oversight by medical practitioners.

(2)

Overeating can be seen as a transgression against life, because overweight is a liability against it. Of course, controlling one's weight is ever a challenge. Devotion to Life can be, not only inspiring, but demanding.

(3)

Smoking is recreational suicide and an offense against our lives.

DRUGS

Life owes much of its modern stability and longevity to drugs. Their curative and palliative effects help enable the stability and enjoyment of our lives, so we may more readily see life's value and perceive its perfectibility.

(2)

When a drug is illegal, it supports an illegal pharmaceutical business. The way to terminate these businesses, along with their harmful effects, is to make all drugs legal. The ones that can only be administered with a doctor's guidance should, of course, be available only by prescription. I also believe that all drugs should be free to the needy – not as a burden to pharmaceutical companies, which must make a profit to support research and attract investment, but as a support service provided by a life-devoted society.

YOUTH & AGING

While we are all at various stages of our lives, in a mass media society, we all share the same information. So the idea of one generation being a complete stranger to another one has become illogical. Today's truer differences among people of different ages are due more to experience and what we select to participate in. Of course, the latter ever allows youth and age to have likes that distinguish them to a healthy degree.

(2)

By the time we become adults, we each contain within ourselves, in our rapidly changing panoply of daily events, a number of generations. In fact, in a world of widely shared content, even the young soon have the experiences of more than one generation within them.

(3)

When we stay up to date, we are ever mentally new. While the young may be innovative in ways that do not require or are not constrained by a great deal of information, the old can be innovative is ways that do or do not require more knowledge and experience.

(4)

The young judge more with the hopes of who they intend to be, while the old judge more with the hopes of who they have become.

(5)

Since the past becomes, inalterably, part of each of our lives, we best adjust by appreciating the good aspects, making peace with the bad ones, and acknowledging that our present lives are the result of both. The way to make peace with the past is to appreciate the present, which exists because of the past.

(6)

The only way to improve the future is to alter the present.

(7)

I'm not even as young as I was yesterday, but I'm younger than I'll be tomorrow.

(8)

Life welcomes us first and time in its time.

(9)

As our mileage increases, the road seems to jostle us more. Be of good cheer. Every age, perceived with life's gifts, is a blessing.

(10)

While there is information everywhere, insight, imagination, sensitivity, humanity and achievement of every kind stand out. We can't afford to be particular about genius; it remains rare and invaluable at any age.

NATURE

To the extent that we appreciate nature, we are inspired to praise it and its Origin; to the extent that we demean it, we are doomed to mock it and its Origin.

WILDERNESS

The wilderness is a shrine of unspoiled Creation.

NATURAL SPIRITS

If moderns ever wish, they can imagine as many perfectly natural spirits as the ancients had mythical ones, by letting the personalization of them spring, in the manner of frankly fictional characters, from the elements and attributes of nature. Then we might say, with some credibility, "May the spirit of the breezes blow well on your sailboat."

Since, however, we value most the elements and attributes themselves, we think there is little need for and likely will never be the desire to create such playful personifications, except perhaps for the amusement of children, because they do resonate of the ignorance rational people have often been the victims of and they tend to remove us from the natural miracles that actually exist. Considering what is required to create a breeze, who should not be happy to know it is at the service of your sail?

POLLUTION

A period of pollution would be expected in a naturally evolving society, particularly in the early stages of industrialization, but we would also expect a reduction of it with the refinement of technology. This course seems to be the one we're on. If we falter, nature will, I think, instruct us, with lessons harsh as we provoke, that it will not go tamely to its grave. It already issues perfectly natural warnings with overheated days and stronger storms. We will either mend our ways or may see such catastrophic effects, including enormous reductions in population, that we will be compelled to revert to a largely preindustrial way of life, where essential services are supplied with minimal electric generation by nonpolluting technologies.

(2)

Clean water comes into the home and clean water could leave it. All we need is a complete water purification system in each home.

(3)

I fear for the health of people who jog in smog.

CITY ENVIRONMENTS

Living beings are most at home in living environments. So, while a great city can be a wonder, we ought to inhabit environments that are akin to the ones we evolved in, so our cityscapes should provide, for example, a plentitude of plants, clean waterways, and fresh air. Then we can feel that we are still part of nature, not uprooted in barren roosts and adrift in over-cemented habitats.

(2)

Cities are inevitable and actually rejuvenating for the countryside; for example, the migration to the city has enabled the reforestation of what was once farmland.

(3)

We should separate walkways and "driveways" with more than curbs. People and moving steel don't mix well.

SAFETY

To protect life, we ought to do our best to be careful, read danger signs and remove the causes of accidents.

MAINTENANCE

The better we care for our personal property, the better we enable our individual lives, and the better we care for our shared property, which is part of humanity's worldwide heritage, the better we enable our mutual lives.

THE OTHER SIDE OF THE ARGUMENT

To help magnify the importance of the conversion from reverence for death to Reverence for Life, let us imagine a world where the change has already taken place. We now have a widespread preponderance of Reverence for Life, are glad to have a share in it, take good care of it, enjoy the fulfillment we can, and view it as a complete gift.

Now, for some reason, the world decides to change back to the way it is today. We once again have a preponderance of irreverence for life, are frequently sorry to be alive, abuse life, let our fulfillments wither every day, and hope for a much better life after we have passed through this vale of tears.

Can you imagine what could precipitate such a bleak and self-punishing reversion? It would, I think, have to be even more catastrophic than the barbarian conquest of the Roman Empire, which precipitated retreat of the benumbed world into the chanting Dark Ages. In fact, we ask if the change will be such a welcome improvement it will make us more secure against a reversion to barbarism than we have ever been and perhaps immune to the threat.

LOVING LIFE

We love life because we appreciate its multitudinous blessings, which, we know, invite our care, enable the fulfillment of our finest possibilities, and are the wondrous gift of Creation.

A FRIEND OF LIFE

Be a friend of life. Have faith in it. Devote yourself to it. Care for it. Do your best with it. Express your gratitude for it. And be content with its apparent limits.

HOPE

The hopes of humanity are coincident with the potential of life, sanctified.

(2)

A philosophy founded on life enables the preservation, fulfillment, and joy of life. It also enables intelligent spirituality.

(3)

Life is, by definition, hopeful. So let us hope.

The End

Printed in the United States
108849LV00002B/163/P

9 780977 420681